Satesh Bidaisee
Joav Merrick
Editors

Public Health

Intersection of Health, Humans, Animals and the Environment

DOI: https://doi.org/10.52305/JZIZ3286

Library of Congress Cataloging-in-Publication Data

ISBN: 979-8-89530-443-3 (Softcover)
ISBN: 979-8-89530-489-1 (eBook)

Published by Nova Science Publishers, Inc. † New York

Contents

Preface

In this book we have tried to examine health from the integrated perspective of humans, animals and the environment to explore human health across selected issues. Content areas engage research, analysis and interpretation of selected topics from the SARS CoV 2 pandemic, social determinants of health, aging and chronic burdens, infectious and zoonotic diseases, food safety, ecology and conservation. The progression of health from a physical, mental and social context is applied within and across each selected topic with emphasis on the challenges and solutions involved. From the inquisition of the researcher to the interest of a student and the intrigue of the public, the narratives presented will engage the knowledge and application of prepared and presented human, animal and environmental intersection.

Section one: Introduction

Chapter 1

The intersection of health, humans, animals and the environment

Satesh Bidaisee[1,*], DVM, MSPH, EdD
and Joav Merrick[2-5], MD, MMedSci, DMSc

[1]Public Health and Preventive Medicine, St George's University, Grenada, West Indies
[2]National Institute of Child Health and Human Development, Jerusalem, Israel
[3]Department of Pediatrics, Mt Scopus Campus, Hadassah Medical Center and Faculty of Medicine, Hebrew University of Jerusalem, Jerusalem, Israel
[4]Kentucky Children's Hospital, University of Kentucky, Lexington, Kentucky, United States of America
[5]Center for Healthy Development, School of Public Health, Georgia State University, Atlanta, United States of America

Abstract

In this book we have tried to examine health from the integrated perspective of humans, animals and the environment to explore human health across selected issues. Content areas engage research, analysis and interpretation of selected topics from the SARS CoV 2 pandemic, social determinants of health, aging and chronic burdens, infectious and zoonotic diseases, food safety, ecology and conservation. The progression of health from a physical, mental and social context is applied within and across each selected topic with emphasis on the challenges and solutions involved. From the inquisition of the researcher to the interest of a student and the intrigue of the public, the narratives presented will engage the knowledge and application of prepared and presented human, animal and environmental intersection.

* ***Correspondence:*** Satesh Bidaisee, Professor, Public Health and Preventive Medicine, St George's University, Grenada, West Indies. Email: sbidaisee@sgu.edu

In: Public Health: Intersection of Health, Humans, Animals and the Environment
Editors: Satesh Bidaisee and Joav Merrick
ISBN: 979-8-89530-443-3

Introduction

In this book we wanted to conduct an examination of health from the integrated perspective of humans, animals and the environment to explore human health across selected issues. Content areas engage research, analysis and interpretation of selected topics from the SARS CoV 2 pandemic, social determinants of health, aging and chronic burdens, infectious and zoonotic diseases, food safety, ecology and conservation.

Everyone can relate, review, refer and recommend the examination of each topic through its background, literature, significance, methodology, discussion and recommendations. The reader will engage each chapter from an academic, governance and civil society perspective towards understanding the causal nature of health as well as the integrated approach to its management.

The progression of health from a physical, mental and social context is applied within and across each selected topic with emphasis on the challenges and solutions involved. From the inquisition of the researcher to the interest of a student and the intrigue of the public, the narratives presented will engage the knowledge and application of prepared and presented human, animal and environmental intersection.

Crime and violence are a growing public health concern, and an intersection perspective is the relationship between animal abuse and tendencies of human criminality. The issue of youth driven animal abuse is imperative to analyze with respect to public health, because human violence is a prevalent issue regarding increasing episodes of domestic violence, mass murders or serial killings. Animal abuse is defined as "all socially unacceptable behavior that intentionally causes unnecessary pain, suffering and/or death to an animal" (1). Having the capability to study this relationship through the analysis of previous literature will help to support the efforts in decreasing the numbers of school shootings, domestic abuse, and murders through raising awareness of this distinct connection and the initial signs of violent tendencies in developing children. Typical acts of violence, such as school shootings, homicides and domestic violence between romantic partners are typically foreshadowed in the form of animal or domestic pet abuse when those same individuals are going through childhood and adolescence. Being aware of this relationship can be an exponential help towards the public safety of many communities that may get targeted due to previous emotional trauma a child was subjected to. Analysis of events can be divided into three pertinent sections: The acts of animal abuse carried out by children, acts of animal abuse that are witnessed by children and acts of animal abuse that occur in the context of domestic violence. It is important to analyze the foundations of the evolution of these behaviors to get afflicted individuals the appropriate treatment needed to prevent their violent behaviors from progressing further. It has been studied and determined that individuals growing up with a tendency to harm innocent creatures tend to grow up and expand their violent tendencies, projecting them onto their peers and other individuals (2). Any aggressive act directed towards an animal is an early indicator of future psychopathology. If this is unrecognized and left without treatment, then it can escalate into more detrimental acts (3). On the topic of more severe acts that can manifest from these foundations, many serial killers were known to have harmed animals as children, examples include Jeffrey Dahmer, Ted Bundy and John Wayne Gacy (4).

Previous studies have examined a distinct relationship between the formation of interest in the abuse of pets and small wild animals and the evolution of that abusive tendency towards other humans as adults. Animal abuse has also been determined to be a precursor to interpersonal violence, this relationship having been termed the "Link" (1). The psychology behind a child's curiosities indicates that there are underlying issues that push them to inflict harm on a creature that is easily attainable and defenseless. The importance of noticing and understanding the origins of these tendencies is so that these children and adolescents receive the necessary therapy and care to properly process their emotions and thoughts, prior to having these tendencies spiral out of control (2).

Recommendations can be made regarding the outcome of this research so that individuals know what habits to look out for, to recognize early signs of animal abuse in children and adolescents, and therefore be able to provide them with appropriate counseling and education in to supplement their psychological development. Recognizing key patterns is imperative in decreasing the risk of future domestic violence or general tendencies towards violence. Further research can also be made to marginalize further influences on this behavioral trend, such as changing the research focus towards victims of domestic abuse who have also witnessed their abuser commit animal cruelty. This shift in focus regarding the research topic will give more insight and more data regarding an adult performing animal abuse alongside domestic violence.

References

(1) Beirne P. From animal abuse to interhuman violence? A critical review of the progression thesis. Society Animals 2004;12(1):39-65.
(2) Lockwood R. Animal cruelty and human violence: The veterinarian's role in making the connection-The American experience. Can Vet J 2000;41(11):876-8.
(3) Tiplady C. Animal abuse: Helping animals and people. Wallingford, Oxfordshire, United Kingdom: CABI Books, 2013.
(4) Flynn C. Examining the links between animal abuse and human violence. Crime Law Soc Change 2011;55(5):453-68.

Section two: Public health issues

Chapter 2

Impact of the COVID-19 pandemic on social isolation and the mental health of adolescents in the United States: A narrative review

Krisha Arora, MD, MPH
Sara Elzibak, BSc
and Satesh Bidaisee*, DVM, MSPH, EdD
Public Health and Preventive Medicine, St George's University, Grenada, West Indies

Abstract

The focus of this narrative review was the emotional distress experienced by adolescents as a result of the COVID-19 pandemic. COVID-19 introduced new social isolation regulations and a state of uncertainty, which affected people in various ways. Adolescents between the ages of 10 and 19 years were a vulnerable population due to their ongoing physical and social development. School closures contributed a great deal to students' unusual feelings of fear, boredom, frustration and anxiety. A literature search was conducted using Google Scholar and PubMed databases to assess the impact of COVID-19 on social isolation and mental health of adolescents. After keywords were used to break down the topic, 122 articles were discovered from the years 2019 to 2021. Next, titles and abstracts were screened, duplicates were removed, and eligibility criteria applied. Finally, the screening procedure produced 21 articles, which were analyzed for this review. The widely observed effects of the pandemic were categorized as anxiety, depression, post-traumatic stress symptoms (PTSS), domestic violence, and worsening

* *Correspondence:* Satesh Bidaisee, Professor, Public Health and Preventive Medicine, St George's University, Grenada, West Indies. Email: sbidaisee@sgu.edu

In: Public Health: Intersection of Health, Humans, Animals and the Environment
Editors: Satesh Bidaisee and Joav Merrick
ISBN: 979-8-89530-443-3

pre-existing mental health conditions. Many articles revealed a combination of the aforementioned outcomes experienced by adolescents who did not previously encounter them prior to the pandemic. COVID-19 social isolation negatively impacted adolescents worldwide, which made it essential to provide the support required for their healthier futures.

Introduction

Mental health forms a significant part of overall well-being; as adolescents transition to adulthood, they gain developmental experience allowing them to balance the psychological and physical components of health (1). In response to the COVID-19 pandemic's social isolation mandate, adolescents experienced a break from routine and sudden increased levels of stress, panic, frustration, loneliness, and anxiety, which led to an overall decline in mental health (1-14). Aggravation of previous disorders and/or development of future disorders were highly probable in vulnerable adolescents who were suffering from adverse childhood experiences (ACEs) during the pandemic (15). Mental illness largely impacts the short- and long-term social and physical well-being of adolescents (2). This effect is a public health concern, because not only a small number of individuals, but a large component of the population had been affected. Additionally, with proper education and resources, adolescents could be provided with the emotional tools to aid them during social isolation and potentially prevent future psychological complications (1, 2, 14, 15).

Adolescents, generally are at risk of novel and worsening mental illnesses encompassing behavioral disorders, attention-deficit/hyperactivity disorder, anxiety and depression (1, 2, 5, 7, 9, 10, 13). For most adolescents, schools served as a "safety net resource" away from home; however, with school closures, they were deprived of access to those as well as in-school mental health services, which led to deteriorating family dynamics (2-5, 7, 8, 11, 16). Pre-COVID-19 data showed more than one-third of adolescents in the United States sought mental health services in school settings (16). This clarifies the struggles faced by adolescents who were unable to receive the psychological care they needed and those whose ongoing care was disrupted during the pandemic (2, 6, 7).

The purpose of this narrative review was to analyze whether COVID-19-related social isolation mandated in the United States has negatively affected adolescents' mental health. In the past, research has been conducted to study large disease outbreaks, such as severe acute respiratory syndrome, Middle

East respiratory syndrome and Ebola concerning their emotional toll on adults; however, those studies excluded information on children and adolescents (1, 8). Although recent studies have concluded the need for social isolation practices to halt the transmission of COVID-19, sufficient research is not yet available on social isolation impact on adolescents in the United States (2, 9).

Our review

This narrative review was conducted to assess the impact of COVID-19 social isolation on the mental health of adolescents.

After the research topic was chosen, relevant articles were obtained from the years 2019 to 2021, using Google Scholar and PubMed databases. Specific keywords which were utilized during the search consisted of: COVID-19 (coronavirus disease-2019, COVID, coronavirus), pandemic, mental health (psychiatric issues, anxiety, depression, stress, behavioral health issues), adolescents (young adults, children), impact (affect, effect), social isolation (isolation, quarantine, social distancing, lockdown) and United States.

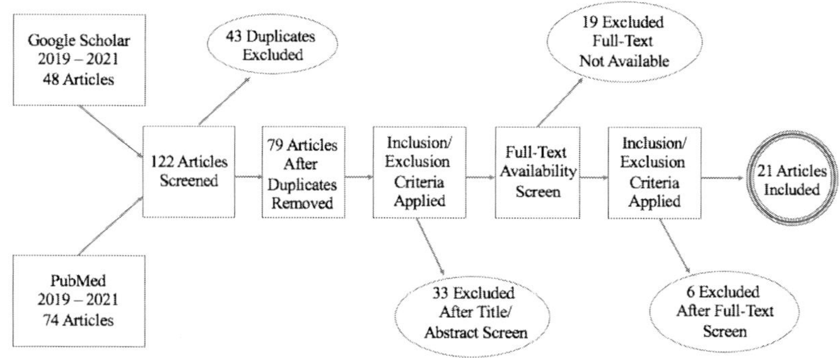

Figure 1. Article selection process.

A total of 122 articles were screened, most of which did not contain sufficient information about the topic upon screening their titles and abstracts. Next, duplicate articles were removed, resulting in 79 articles. Then, eligibility criteria were applied to the articles. The inclusion criteria consisted of articles about adolescents ages 10-19 years, based on social isolation during COVID-19, and the mental health effects of the pandemic. Exclusion criteria were articles based solely on countries other than the United States and/or China, articles in languages other than English, articles written prior to 2018, and those about adolescents diagnosed with COVID-19. Many of the relevant articles were not found in full-text formats, resulting in an ultimate total of 21 articles remaining to conduct the review, as seen in Figure 1.

The selected articles were analyzed for all information pertaining to the ways in which COVID-19 lockdown and social isolation have affected adolescents. Specifically, changes in levels of stress, panic, anxiety and depression in relation to social isolation were noted. Changes in family dynamics leading to more negative interpersonal relationships among family members were also evident in addition to negative effects on pre-existing mental health conditions.

Our findings

After the relevant information was extracted from the selected articles, it was organized by outcome and summarized in Table 1. Adolescent mental health was noted to be impacted greatly by social isolation with respect to certain sequelae specifically, when compared to data prior to the pandemic. Increased anxiety levels were noted in 18 articles, while increased depression symptoms were noted in 17 articles (1-4, 6-10, 13-15, 17-22). PTSS in addition to general stress had also increased post-social isolation in 16 of the articles (1-3, 6-9, 11, 12, 14, 17-22). Domestic violence rates increased during the pandemic as seen in eight of the 21 articles (2-4, 6-8, 14, 22). Finally, adolescents with pre-existing mental health conditions claimed to experience a negative impact during social isolation in 16 of the 21 articles (1-8, 11-13, 15, 18, 20-22). Two-thirds of the articles emphasized the importance of schools and education centers as major providers of mental health services for adolescents who were greatly impacted by social isolation and virtual education mandates (1, 2, 4, 5, 7, 8, 11, 13, 14, 16, 17, 19, 21, 22).

Table 1. Number of articles documenting each condition during the COVID-19 pandemic

COVID-19 outcome	Number of articles	References
Anxiety	18	(1-4, 6-10, 13-15, 17-22)
Depression	17	(1-4, 6-10, 12, 15, 17-22)
Post-traumatic stress symptoms (PTSS)	16	(1-3, 6-9, 11, 12, 14, 17-22)
Domestic violence	8	(2-4, 6-8, 14, 22)
Worsening pre-existing mental health conditions	16	(1-8, 11-13, 15, 18, 20-22)

Discussion

COVID-19-related social isolation led to heightened emotional distress, boredom, frustration, fear, and loneliness in adolescents due to altered routines and long breaks from their norm (1, 2, 4, 7-9, 11, 12, 14). The vast change in emotions and uncertainty regarding the virus and pandemic introduced confusion and stress into their lives, ultimately presenting as newly diagnosed and/or worsening psychiatric conditions including anxiety and depression (1-15, 17-22). A deterioration in mental health occurred, which was difficult to overcome while isolated from family, friends, health professionals, education professionals, counselors and others not found in individuals' households during the pandemic.

Social isolation regulations, specifically those regarding virtual education and virtual workplace, caused adolescents and their families to spend a lot more time together than before the pandemic started. Both positive and negative outcomes resulted from this change in the daily environment. Positive outcomes that surfaced were building better interpersonal relationships with siblings and caregivers, in addition to having more free time to enjoy family activities. On the other hand, there were many cases of negative consequences for spending more time together with family while experiencing the stressors of COVID-19. Some studies found that domestic violence increased throughout the pandemic, as families were living together under stressful conditions; parents had to work from home while partaking in full-time childcare (2-4, 6-8, 14, 22). Also, many students' major support system for mental health queries was oftentimes a figure in their schools (2-5, 7, 8, 11, 16). Therefore, with school closures, it became difficult for students to trust another person with such delicate information about their personal lives, so they no longer received supportive counseling.

Adults, elderly, children and adolescents all experienced changes in their established lives throughout the COVID-19 pandemic; however, children and adolescents were a far more vulnerable population during the uncertain times. There was a level of helplessness beyond the understanding of many, which in addition to hormonal and developmental changes made the situation tougher for adolescents to overcome. Even though adolescents were less likely to suffer from symptoms of COVID-19, they were one of the most vulnerable groups to experience the negative impact of the pandemic. Stress, fear, and other ACEs associated with social withdrawal triggered neuroinflammation, leading to an increased predisposition to psychopathological illnesses in adulthood, including depression, anxiety, psychosis, and addiction (2, 23).

While the impact of social isolation on adolescent mental health was a negative one on a global level, there seemed to be a lack in data for the United States as opposed to China, Europe, and India. Studies exist from around the world, including the United States with a focus on adults' mental health during previous pandemics, however, information is scarce for children and adolescents. One limitation of the studies conducted to yield the data was the response rate to surveys, questionnaires, and assessments. Additionally, the accuracy of responses was another area of concern due to self-reporting by adolescents introducing some reporting bias. Another limitation to finding more articles and/or studies on the topic was the fact that the COVID-19 pandemic is still very recent. Therefore, at this time it is not possible to document the long-term effects of the pandemic on adolescent mental health as not much time has passed since the start of the pandemic (21).

People worldwide have been impacted by the COVID-19 pandemic, especially due to the mandated social isolation and school/work closures. The major negative outcomes on the mental health of adolescents ranging from anxiety and PTSS to depression and domestic violence stemmed from the confusion, stress, and uncertainties revolving around the isolation restrictions set forth by the government. This is a public health concern, because the resulting behavioral health issues were not short-lived and could be prevented with suitable strategies and better understanding of the situation.

The focus of public health is to avoid, prevent, and adapt to health-related issues on a community and/or population level. A pandemic not only affects individuals but the entire exposed population. Adolescents comprise a large, vulnerable, and important part of the population as a whole; their well-being creates better outcomes for the future of the community. In addition to the short-term behavioral health issues faced by adolescents, the long-term impacts are of major importance, especially because they could potentially be overlooked for years. Ongoing neurodevelopment, along with exposure to increased stress and ACEs during the pandemic, make adolescents more susceptible to the long-term physiological and psychological consequences of social isolation (2). In the future, there would also be a large impact on the health systems if the long-term effects follow adolescents into adulthood leading to a vast majority of them needing health services; therefore, it should be treated as a public health concern (23). Public health officials, government officials, health authorities, communities, organizations, educators, and families should all work together to provide vulnerable populations with the tools and strategies necessary to avoid severe psychiatric consequences and take better care of themselves in difficult situations.

Since literature verifies the negative outcomes of the pandemic experienced by adolescents, recommendations for prevention strategies to deal with the consequences are warranted. Therefore, they should be provided with dedicated education and strategies on how to tackle situations like the COVID-19 pandemic along with its regulations and uncertainties. With an instructional plan focused on behavioral health and well-being, adolescents would be better equipped to avoid the suffering caused by the pandemic. Those who relied solely on certain environments and people to help them deal with their problems were at a disadvantage during quarantine and isolation. However, they could learn to cope with those issues in the future by being taught the skills and techniques required to care for themselves in difficult circumstances.

Recommendations

The following recommendations would be beneficial for adolescents during a pandemic. While schools are closed, students should still follow a set schedule and partake in group activities, virtually or with those in their households. Social interactions should be maintained as much as possible so that healthy relationships may continue to form and not disappear (9, 12). Physical activity is also very important for caregivers and educators to promote while students are learning virtually. It would be ideal to maintain and mirror most of the adolescents' activity level at home to that while actually attending school.

Another recommendation is pandemic education; during virtual classes, teachers should spend time explaining facts about the pandemic. With more education on factual information regarding the background and updates on the virus and pandemic, adolescents would gain a better understanding of the situation which would ultimately diminish stress related to uncertainty (1, 10, 21, 22). In addition to pandemic education, psychosocial education should also be expanded in the curriculum. Positive mental health should be promoted with exercises and interventions available in-person and virtually/digitally; with these services and support, positive outcomes will be achieved (17-22). While having psychological services available in virtual classrooms, they should also be expanded into healthcare as "general pandemic healthcare," which would emphasize the importance of mental health as a regular component of primary care during a pandemic (12).

Conclusion

As the COVID-19 pandemic continues, many changes will have to be made globally, to support the wellbeing of adolescents and the general population. Policy makers, educators, researchers, administrators, and community members must work together to reduce pandemic-related negative effects on adolescents. Future research is warranted in order to document the long-term effects of social isolation during the COVID-19 pandemic in addition to the evolving mental health care needs of adolescents in the United States.

References

(1) Shah K, Mann S, Singh R, Bangar R, Kulkarni R. Impact of COVID-19 on the mental health of children and adolescents. Cureus 2020;12(8):e10051.

(2) de Figueiredo CS, Sandre PC, Portugal LCL, Mazala-de-Oliveira T, Chagas LS, Raony I, et al. COVID-19 pandemic impact on children and adolescents' mental health: Biological, environmental, and social factors. Prog Neuropsychopharmacol Biol Psychiatry 2021;106:110171.

(3) de Miranda DM, da Silva Athanasio B, Sena Oliveira AC, Simoes-E-Silva AC. How is COVID-19 pandemic impacting mental health of children and adolescents?. Int J Disaster Risk Reduct 2020;51:101845.

(4) Ghosh R, Dubey MJ, Chatterjee S, Dubey S. Impact of COVID-19 on children: Special focus on the psychosocial aspect. Minerva Pediatr 2020;72(3):226-35.

(5) Golberstein E, Wen H, Miller BF. Coronavirus disease 2019 (COVID-19) and mental health for children and adolescents. JAMA Pediatr 2020;174(9):819-20.

(6) Guessoum SB, et al. Adolescent psychiatric disorders during the COVID-19 pandemic and lockdown. Psychiatry Res 2020;291:113264.

(7) Khan KS, Mamun MA, Griffiths MD, Ullah I. The mental health impact of the COVID-19 pandemic across different cohorts. Int J Ment Health Addict 2022;20(1):380-6.

(8) Lee J. Mental health effects of school closures during COVID-19. Lancet Child Adolesc Health 2020;4(6):421.

(9) Loades ME, et al. Rapid systematic review: The impact of social isolation and loneliness on the mental health of children and adolescents in the context of COVID-19. J Am Acad Child Adolesc Psychiatry 2020;59(11):1218-39.

(10) Oosterhoff B, Palmer CA, Wilson J, Shook N. Adolescents' motivations to engage in social distancing during the COVID-19 pandemic: associations with mental and social health. J Adolesc Health 2020;67(2):179-85.

(11) Patel M, Raphael JL. Acute-on-chronic stress in the time of COVID-19: Assessment considerations for vulnerable youth populations. Pediatr Res 2020;88(6):827-8.

(12) Pfefferbaum B, North CS. Mental health and the Covid-19 pandemic. N Engl J Med 2020;383(6):510-2.

(13) Smirni P, Lavanco G, Smirni D. Anxiety in Older Adolescents at the Time of COVID-19. J Clin Med 2020;9(10):3064.
(14) Wang G, Zhang Y, Zhao J, Zhang J, Jiang F. Mitigate the effects of home confinement on children during the COVID-19 outbreak. Lancet 2020;395 (10228):945-7.
(15) Wagner KD. Addressing the experience of children and adolescents during the COVID-19 pandemic. J Clin Psychiatry 2020;81(3):20ed13394.
(16) Ali MM, West K, Teich JL, Lynch S, Mutter R, Dubenitz J. Utilization of mental health services in educational setting by adolescents in the United States. J Sch Health 2019;89(5):393-401.
(17) Fruehwirth JC, Biswas S, Perreira KM. The Covid-19 pandemic and mental health of first-year college students: Examining the effect of Covid-19 stressors using longitudinal data. PLoS One 2021;16(3):e0247999.
(18) Jones EA, Mitra AK, Bhuiyan AR. Impact of COVID-19 on mental health in adolescents: a systematic review. Int J Environ Res Public Health 2021;18(5):2470.
(19) Gazmararian J, Weingart R, Campbell K, Cronin T, Ashta J. Impact of COVID-19 Pandemic on the Mental Health of Students From 2 Semi-Rural High Schools in Georgia. J Sch Health 2021;91(5):356-69.
(20) Gotlib IH, Borchers LR, Chahal R, Gifuni AJ, Teresi GI, Ho TC. Early life stress predicts depressive symptoms in adolescents during the COVID-19 pandemic: The mediating role of perceived stress. Front Psychol 2021;11:603748.
(21) Meherali S, Punjani N, Louie-Poon S, Rahim KA, Das JK, Salam RA, et al. Mental health of children and adolescents amidst CoViD-19 and past pandemics: a rapid systematic review. Int J Environ Res Public Health 2021;18(7):3432.
(22) Scott SR, Rivera KM, Rushing E, Manczak EM, Rozek CS, Doom JR. “I hate this”: A qualitative analysis of adolescents' self-reported challenges during the COVID-19 pandemic. J Adolesc Health 2021;68(2):262-9.
(23) Henderson MD, Schmus CJ, McDonald CC, Irving SY. The COVID-19 pandemic and the impact on child mental health: A socio-ecological perspective. Pediatr Nurs 2020;46(6):267-90.

Chapter 3

Recognition of intimate partner violence using telemedicine during the COVID-19 pandemic

Mandeep Grewal, MD, MPH
Sara Elzibak, BSc
and Satesh Bidaisee*, DVM, MSPH, EdD
Public Health and Preventive Medicine, St George's University, Grenada, West Indies

Abstract

In response to the recent COVID-19 pandemic, healthcare providers have shifted to telemedicine consultations that unwillingly produced additional distress in intimate partner violence (IPV) victims. This chapter aims to raise awareness of increasing coping difficulty in IPV victims, help clinicians recognize its red flags during telemedicine, and compile strategies on safe and effective communication. A literature review was conducted on Google Scholar and PubMed. After reviewing eighteen selected articles, the signs of potential IPV have been summarized to help clinicians identify the victims, ask appropriate questions, and strategize their approach. The multiple studies showed worldwide uptrends in IPV during this pandemic. All the studies supported for universal education while screening all female patients, a harm reduction approach with ensuring confidentiality, and empowering the victims with appropriate referrals. The patient-centered approach with more details on red flags of IPV, questions for victims, and strategies for professionals have been summarized in this paper. Despite the challenges faced by physicians during telemedicine, the skills

---

* ***Correspondence:*** Satesh Bidaisee, Professor, Public Health and Preventive Medicine, St George's University, Grenada, West Indies. Email: sbidaisee@sgu.edu

In: Public Health: Intersection of Health, Humans, Animals and the Environment
Editors: Satesh Bidaisee and Joav Merrick
ISBN: 979-8-89530-443-3

acquired in previous experiences should be continued when interacting with IPV victims. The ability to recognize the non-verbal findings of IPV, and the skill to examine victims in a compassionate and sensitive matter needs an update during these emerging crises.

Introduction

According to the United Nations (UN), "domestic violence" or "intimate partner violence (IPV)" can be defined, "as a pattern of behavior in a relationship that is used to gain or maintain power and control over an intimate partner" (1). Spousal abuse is another term commonly used to represent such behavior by a partner, too (2). It is crucial to understand that violence can take any form – physical, sexual, emotional, economic, psychological, or digital (1, 3). Multiple studies have identified the following risk factors of IPV e.g., lower socioeconomic status, low education level, inadequate social support system, younger age, unintended pregnancy, financial dependency on partner, unemployment, care and home schooling of children, increased drug and alcohol use, and the previous history of victimization experiences (4-6).

IPV is reported to be the most common cause of nonfatal injury to women worldwide (4, 7). Before COVID-19 pandemic, the global-reported rate of IPV was about 1 in 3 women (8). It is undeniable that IPV causes significant morbidity on women's health by affecting their mental wellness (7, 9). During this pandemic, the majority of primary healthcare providers have shifted to telemedicine for the public health safety (10). It has led to increased distress on victims and survivors of IPV to seek help when their perpetrator of violence is present in the same environment during isolation at home (11, 12). Furthermore, resulting in oppression and neglection of victims/survivors (13). This pandemic exacerbated public health issues in IPV victims, including lack of privacy, emotional toll of isolation, and access to safe places in community that resulted in negatively impacting their coping strategies (14). It is also not unusual when a pandemic has brought in shadow crisis to our health (15, 16). Historically, the uptrends in rates of violence against women have also been witnessed during disasters like hurricanes, earthquakes, and Great Recession of 2007-2009 that negatively impacted women health and well-being (3, 5, 17).

This chapter aims to raise awareness among all healthcare providers of increasing IPV victims' coping difficulty during this pandemic, help them recognize red flags of IPV when using telemedicine, and provide tools in preparing themselves to tackle this "shadow pandemic."

Our review

This narrative review is secondary research, and a non-experimental study designed to evaluate literature describing increase in IPV during COVID-19 pandemic and compile strategies for its early recognition during telemedicine visits. The Google Scholar search was conducted using advanced search feature with all the following words, "domestic violence," "SARS2," "disrupted social structure," "family dynamics," "family structure," "intimate partner violence," "coronavirus," "SARS CoV 2"; with the exact phrase, "COVID-19"; and limited the articles publication since 1 Jan 2020. This search yielded 396 articles. The PubMed search was also conducted between 4 February 2021 and 3 March 2021 using MeSH terms "spouse abuse" OR "domestic violence" OR "intimate partner violence" AND "telemedicine"; and limiting the articles publication since 1 Jan 2020. This search yielded thirty-three (33) articles. In total, both the searches produced four hundred and twenty-nine (429) articles that were screened manually. There were seventy-two (72) duplicates identified. The most common reason for excluding articles was their irrelevance to IPV because the majority were focusing on COVID-19 pandemic's impact in other fields like pediatrics, education, economy, employment, and technology. Fifty-five (55) articles were assessed by obtaining full-text and thirty-seven (37) were removed because their focus was different than the strategies for healthcare providers or public health officials to deal with IPV during telemedicine. A detailed study flow chart can be seen in Figure 1. The total of 18 articles were selected and included in this review study.

Inclusion criteria: women aged >18 years, articles discussing domestic violence, spouse abuse, or intimate partner violence and women health or mental health during COVID-19 pandemic. The articles were not limited to any geographical location because this virus has spread internationally. The full-text articles were obtained and reviewed to be included in this review only if they discussed strategies for public health officials and healthcare providers to deal with IPV during telemedicine.

Exclusion criteria: articles which enclosed IPV in different patient population such as pediatrics, men, elderly, transgender women, and pregnant women. Populations containing any chronic medical conditions and/or other coexisting disabilities were also excluded from this study. After reviewing the full-text, any articles containing discussion on COVID-19 irrelevant to IPV on women were excluded. A few articles only identified the trends in IPV on women during this pandemic but did not focus on strategies in recognizing or managing them over telemedicine. Articles focusing on online gender-based violence were also excluded.

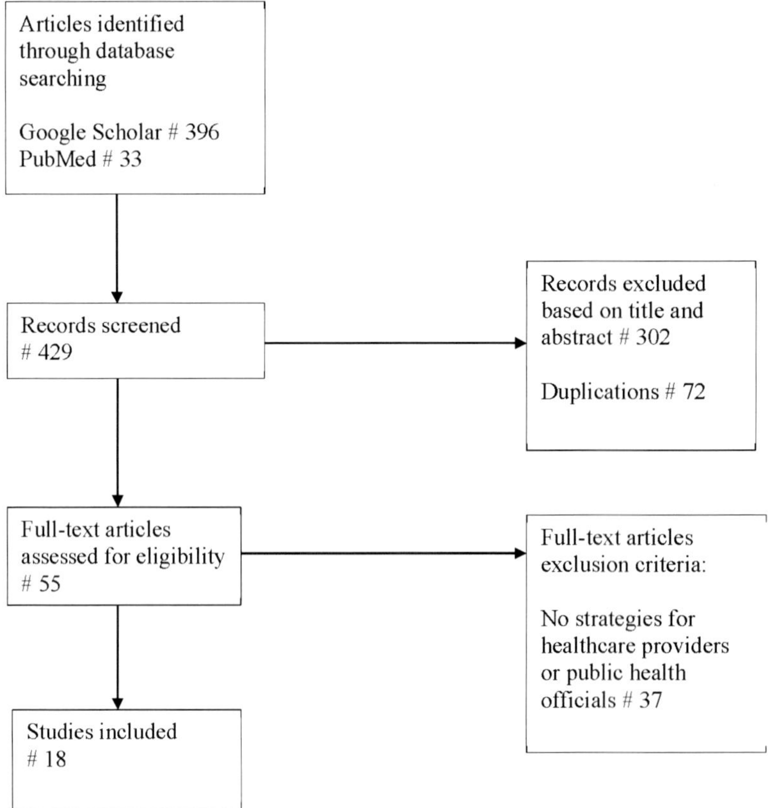

Figure 1. Study flow chart.

The information derived from the literature was reviewed, collated, and provided as a narrative. Analysis of the literature was conducted to present the perspectives from the literature. In this narrative review, the prevalence and trends of change in IPV were compared between different countries. The risk factors for IPV were also compiled from the included studies to help the healthcare providers identify high-risk patient population. An attempt to educate the healthcare service providers was made through this paper by generating the list of questions and strategies that can be discussed with IPV victims during their telemedicine visit. The literature review occurred within the parameters of inclusion and exclusion criteria to inform the selection of literature to be included. Any bias may be associated with the choice of enquiry into the literature based on interest as well as the selection of search words/terms which will influence the results of the literature search and ultimately the review with information bias.

What we found

The United Nations Population Fund (UNFPA) estimated IPV incidence to be 15 million globally for every three months the COVID-19 lockdown continues (10). The 20 major metropolitan cities of USA have been reported to have observed 20% or greater increase in IPV-related calls (11). In USA, the police records from New York City, Alabama, Oregon, San Antonio, Jefferson, Portland, and Texas reported 10-27% increase in IPV during this pandemic (18). There have been a 300% increase in calls to Vancouver, Canada-based IPV crisis line, 150% surge within the first week of stay-at-home measures in UK (5). France also noted 30% increase in police reports regarding IPV during pandemic (4). The Center for Studies and Applied Sciences in Gender, Family, Women, and Adolescent (CSAGA) has reported 624 cases of gender-based violence in the first four months of 2020 compared to 416 in last four months of 2019 in Vietnam (19).

Carrington et al., surveyed 314 service providers in Australia on whether their clients with IPV reported controlling behaviours (20). 135 (43%) service providers have highlighted identifying increase in control and coercion by the abusers in their IPV victims/patients (20). 272 (87%) service providers responded that increased isolation was the most common controlling behavior among patients with IPV, 219 (70%) reported increased sense of vulnerability, 201 (64%) reported the inability to seek outside help, 154 (49%) reported an increased fear of monitoring by the abuser, 147 (47%) reported enhanced surveillance by the abuser, and 118 (38%) reported an increased use of technology to intimidate (20).

Multiple studies reported the non-verbal signs/red flags of potential IPV e.g., “avoidance of eye contact or suspicious traumatic/unexplained injury on the head, neck, or forearms; the patient’s tone of voice including fearful, reluctant to speak, and evasive; wearing long sleeves in summer to hide injuries; low self-esteem; constantly looking up to the partner for minor decisions while the patient is reporting other psychosocial concern; sexual assault; suicide attempt; overdose; vague complaints; delay in seeking care; and repeated emergency room visits" (10, 21-23).

When screening female patients for IPV, healthcare professionals should focus on asking the following suggested questions while aiming to assess the patient's current knowledge on the disease and public health policies, the impact on their gender-based role, the impact on mental health, awareness of the available services, and safety plan. The victims' knowledge on current public health policies is suggested to be assessed to ensure the abusers are not using misinformation to maintain coercive control (12, 20, 24). The division of household labor within a family is suggested to be explored by asking questions, such as, "Who does household chores and takes care of the children/sick/elderly? Who is primarily responsible? How has the pandemic changed their roles at home? What do they think about these changes?" (19, 25). The psychological health impact is suggested to be identified by asking, "Has your mental health changed since the pandemic began? Has there been any conflict in the family since the lockdown?" (19, 20, 26). It is also suggested to ask direct questions when screening IPV victims, such as, "Have you been injured by your partner? Are there weapons in the home, and if so, have you been threatened in any way with them? Have you been threatened to infect with COVID19?" (19, 20, 26). Many authors have recommended providing universal education while screening e.g., "Do you know what services are still being offered in this pandemic? What is offered at each shelter?" (10, 15, 24, 27). The evidence supports asking IPV victims for safety plan e.g., "Do you have access to safe phone/password-protected device or internet services? Who to contact for help when needed? Do you have a safest place within the home or outside?" (12, 15, 22-24). Slakoff et al., (24) has recommended asking the victims if they have established the safe words, such as "we are out of milk" or signals of distress, such as "tucked-thumb and closed-fist hand gesture," "flickering outdoor lights," "opening or closing blinds," and "waving from certain windows," with family, friends, and neighbors. LaPlante et al., (28) discussed key communication skill when interacting with IPV victims and recommended avoiding asking leading questions e.g., "You are not being hurt, are you?"

The evidence has been found in support of making it a routine for healthcare professionals to screen all the women for an IPV and provide universal education as well as share resources, especially when any of the red flags are present (26, 28). It is their responsibility to ensure the transition for patients to telemedicine visits is as smooth as possible and handling the technical difficulties during sessions (20, 22). They recommended not to forget providing reassurance to the patient for their privacy and confidentiality (15, 23, 29). They suggested that practitioners be flexible with scheduling and availability to discuss IPV (28). In order to ensure that the survivor is in a safe space during the telemedicine interviewing, some authors encouraged the use of headphones (22, 28). The healthcare professionals have also been advised to be prepared to address the survivor's safety during the telemedicine interviewing e.g., "While we are in the middle of a session, if you feel threatened by the presence of your spouse/partner, you may indicate it to me using our agreed code/hand gesture or switching to a neutral topic such as weather, menstruation, etc." (11). It is also recommended to review safety practices with the victims, such as "deleting internet browsing history or text messages, saving hotline information under other listings e.g., grocery store or pharmacy, and creating a new, confidential email account for receiving information about resources or communicating with healthcare providers. (30)" During any in-person encounter, the healthcare professionals may use the "No visitor accompanying the patient" policy in COVID-19 to the victims' advantage in cases of suspected IPV (20). In telemedicine interviewing, using only the direct "yes" or "no" questions, such as the STAT screening tool, is recommended (21). Upon the disclosure of IPV, the provider should acknowledge it, listen and validate their feelings, assess the risk of danger, emphasize that violence is never OK, and reassure that help is available (10, 24). The evidence has been found for using online training courses with certification for nurses, other clinic staff, psychotherapists to recognize IPV that must be shared by the healthcare providers (22, 24). The clinicians should be aware that technologists with device security expertise are available and can be provided with remote assistance to IPV survivors (23). They should be aware that there are digital interventions that can help victims make informed decisions, such as myPlan, I-DECIDE, iSafe, M-Health apps tools for developing safe plan. Some apps are password-protected, and some are disguised as news or weather apps for safety (11, 15). When planning and offering the interventions, it is important to explore the victim's support system, assess the risk of immediate danger and make safety plans same as previous in-person visits. However, providing referrals to relevant services and following-up with the victims in the form of personal and secure emails/calls is important (15, 20, 22-24, 26, 27).

Discussion

It is valuable for the healthcare providers to understand the risk factors. They can be incorporated into interviewing questions during telemedicine for the patients. Providers should make it a routine to screen their female patients for IPV during every tele-visit same as the in-person visit, which is in-line with the current USPSTF recommendation of screening all women of childbearing age for IPV (31). There are multiple screening tools that exist in practice e.g., "HITS (Hurt, Insult, Threaten, Scream), STAT (Slapped, Threatened, and Throw), AAS (Abuse Assessment Screen), PVS (Partner Violence Screen), and WAST (Woman Abuse Screen Tool)" (21, 28, 32). These screening tools require least administrative work as they can be self-reported using Information and Communication Technologies (ICT). Alternatively, the providers can have their staff to get the patients screened while triaging. All these screening tools have a common parameter of asking direct questions for assessing the IPV, which is an appropriate therapeutic communication technique in encouraging the expression of patient's true feelings and ideas.

With the emerging of new diseases, perpetrators have also found new methods to threaten, coerce, and take control of their victims e.g., telling her that she can't leave due to restrictions, threatening to infect her with COVID-19, and spreading rumors about the victims having COVID-19 to isolate them further from people (20). The official Governmental orders of wide-spread lockdowns and isolation in this pandemic with new infectious disease have unknowingly increased entrapment, monitoring, and surveillance of IPV victims at their homes (20). The weaponizing of COVID-19 restrictions by perpetrators have further increased the range and intensity of abusive behavior (20). It has impacted victim's mental health with more severe emotional and psychological abuse (19, 20). These public health safety recommendations have instead created additional barriers for some survivors to leave violent relationships (20).

Although widespread "stay-home" orders were put in place keeping in mind the public health safety, it not only precipitated the exposure of vulnerable to IPV but excluded them from healthcare too (5). With many offices being closed immediately following the public health guidelines, victims were left strangulated without appropriate support services for a while. There were limited resources with face-to-face access during COVID-19 pandemic that increased complexity of victim's needs (20). Under such vulnerable and unpredictable circumstances, the public health agencies along with the Government and other stakeholders should take responsibility in communicating with the public in regular and consistent updates about COVID-19 guidelines. This communication can occur at multiple platforms such as newspaper, television, radio, and social media campaigns to prevent abusers from misusing/weaponizing the COVID-19 restrictions and promote sharing of work and mutual support in a time of crisis (19). In response, European countries as well as Canada launched "Signal for Help" and "Safe word" campaigns to inform healthcare workers and patients of gestures, signals, or words to allow for silent request for help during telemedicine (4).

The arrival of this shadow pandemic demands that healthcare professionals including Family Physicians, Psychotherapists, Social Workers, and others involved in the IPV support services to meet this new challenge with innovative solutions in providing their therapeutic support (22). As a service provider, adapting to videoconferencing for telemedicine visits is a solution deemed necessary during this pandemic (20). The patient's right to confidentiality must be protected during these telemedicine visits similar to their in-person visits. The healthcare providers should consider the risks of the IPV victims being monitored by their perpetrators and identify the preferred method of communication with their patients prior to this transition (21).

The healthcare professionals need to ensure that this transition for the patients is as smooth and safe as possible (22). For example, encouraging the patients to use headphones during a telemedicine visit will ensure privacy to the therapeutic conversation (23). It is the duty of professionals to develop remote consultation protocols and train the staff to help recognize IPV (23, 24). The organizations such as IMPACT provide online courses that can be completed at one's own convenience (24). If suspect IPV then ask patient to visit office alone because of no visitor's policy during pandemic can be used for own advantage (20). Providing the IPV victims with a flexible scheduling, longer than usual appointment times, and ensuring to follow-up on their consultations and/or recommendations is an approach increasing effectivity and efficiency of the services (23).

There are concerns of access, cost, and therapeutic viability in telemedicine that need to be addressed for its long-term success (22). Populations with low socio-economic status in developed and developing countries might lack access to suitable technology. With the help of governmental funding, subsidies, and/or rebates, patients and service providers can overcome the cost barrier to video therapy. This pandemic has not only fastened the transition of healthcare services via telemedicine but allowed patients to become acceptable and embrace innovative ways of improved access in healthcare settings. In future, we hope to see expansion of hybrid services consisting of telemedicine with follow-up face-to-face consultations becoming effective at providing access to support services tailored to the IPV survivors needs. We also hope to see more broadened use of technology in the form of smartphone apps and websites with live chat options to help navigate through various service providers securely and efficiently. The other review studies have concluded that digital applications providing screening and safety decision aids were found to be safe, acceptable, accessible by its users, and high satisfaction with these services (27).

Limitations

The risk of information bias and internal validity exists in this study due to its small sample size. It does limit the ability to conduct more detailed analysis in understanding the protective factors that are associated with early recognition of IPV during telemedicine. However, there is a need for conducting further studies and developing evidence-based standard operating protocols when interacting with IPV victims during telemedicine.

Conclusion

The access and delivery of appropriate healthcare services were challenging during COVID-19 restrictions (33). Although there was a negative impact in all patient populations, the IPV victims were suffering increased risk of violence and exclusion due to additional effects of isolation with their abusers (33). The practitioners had to quickly adapt and organize resources during these emerging circumstances. It becomes the responsibility of Governmental as well as public health officials, and healthcare professionals to undertake disaster management planning for IPV. The Governmental and public health officials need to raise awareness among professionals and public on how to use secret words or signals to request for help from home (25, 27). The healthcare professionals not only need to ensure that they are available to their vulnerable populations remotely. They need to ask direct yet open-ended questions to their patients and be able to recognize signs of potential IPV while keeping risk factors in mind, use the restrictions orders to their advantage, and provide safe as well as smoother transition to telemedicine services. All physicians must become comfortable in initiating conversations with female patients in screening them for IPV and educating as well as empowering them by offering information about healthy relationships universally. The innovative ways of utilizing technology such as video therapy to improve access has been found to be welcoming by the patients and resulting in good outcomes when providing therapeutic support for mental health.

References

(1) United Nations. What is domestic abuse? United Nations, 2021. URL: https://www.un.org/en/coronavirus/what-is-domestic-abuse.

(2) Public Health Agency of Canada. Spousal and partner abuse - It can be stopped. Canada.ca. 2009. URL: https://www.canada.ca/en/public-health/services/health-promotion/stop-family-violence/publications/spousal-partner-abuse-stopped.html.

(3) Rauhaus BM, Sibila D, Johnson AF. Addressing the increase of domestic violence and abuse during the COVID-19 pandemic: A need for empathy, care, and social equity in collaborative planning and responses. Am Rev Public Adm 2020;50(6-7):668-74.

(4) Bradley NL, DiPasquale AM, Dillabough K, Schneider PS. Health care practitioners' responsibility to address intimate partner violence related to the COVID-19 pandemic. CMAJ 2020;192(22):E609–10.

(5) Kaukinen C. When stay-at-home orders leave victims unsafe at home: Exploring the risk and consequences of intimate partner violence during the COVID-19 pandemic. Am J Crim Justice 2020;45(4):668–79.

(6) Centers for Disease Control and Prevention. Risk and protective factors, 2021. URL: https://www.cdc.gov/violenceprevention/intimatepartnerviolence/riskprotectivefactors.html.

(7) World Health Organization. Injuries and violence. The facts. The magnitude and causes of injuries. Geneva: WHO, 2014. URL: http://www.who.int/violence_injury_prevention/media/news/2015/Injury_violence_facts_2014/en/.

(8) WHO. Violence against women. Women Stud Abstr 2017. URL: https://www.who.int/news-room/fact-sheets/detail/violence-against-women.

(9) Van Gelder NE, Van Rosmalen-Nooijens KAWL, Ligthart SA, Prins JB, Oertelt-Prigione S, Lagro-Janssen ALM. SAFE: an eHealth intervention for women experiencing intimate partner violence – study protocol for a randomized controlled trial, process evaluation and open feasibility study. BMC Public Health 2020;20(640):1–8.

(10) Hudson LC, Lowenstein EJ, Hoenig LJ. Domestic violence in the coronavirus disease 2019 era: Insights from a survivor. Clin Dermatol 2020;38(6):737–43.

(11) Jarnecke AM, Flanagan JC. Staying safe during COVID-19: How a pandemic can escalate risk for intimate partner violence and what can be done to provide individuals with resources and support. Psychol Trauma 2020;12(S1):S202–4.

(12) Usher K, Bhullar N, Durkin J, Gyamfi N, Jackson D. Family violence and COVID-19: Increased vulnerability and reduced options for support. Int J Ment Health Nurs 2020;29(4):549–52.

(13) Galea S, Merchant RM, Lurie N. The mental health consequences of COVID-19 and physical distancing the need for prevention and early intervention. JAMA Intern Med 2020;180(6):817–8.

(14) Mantler T, Shillington KJ, Davidson CA, Yates J, Irwin JD, Kaschor B, et al. Impacts of COVID-19 on the coping behaviours of Canadian women experiencing intimate partner violence. Glob Soc Welf 2022;1–16.

(15) Emezue C. Digital or digitally delivered responses to domestic and intimate partner violence during COVID-19. JMIR Public Health Surveill 2020;6(3):1–9.

(16) Ravindran S, Shah M. Unintended Consequences of lockdowns: COVID-19 and the shadow pandemic. NBER Work Pap Ser 2020;21(1):1–33.

(17) Kofman YB, Garfin DR. Home is not always a haven: The domestic violence crisis amid the COVID-19 pandemic. Psychol Trauma 2020;12(S1):S199–201.

(18) Boserup B, McKenney M, Elkbuli A. Alarming trends in US domestic violence during the COVID-19 pandemic. Am J Emerg Med 2020;38(12):2753–5.

(19) Hong Giang LT, Thanh Huong NT. CARE rapid gender analysis for COVID-19 Vietnam. ReliefWeb 2020. URL: https://reliefweb.int/report/viet-nam/vietnam-care-rapid-gender-analysis-covid-19-may-2020.

(20) Carrington K, Morley C, Warren S, Harris B, Vitis L, Ball M, et al. The impact of COVID-19 pandemic on domestic and family violence services and clients. Aust J Soc Issues 2021; 56(4):539-58.

(21) Ver C, Garcia C, Bickett A. Intimate partner violence during the COVID-19 pandemic. Am Fam Physician 2021;103(1):6–7.

(22) Simpson S, Richardson L, Pietrabissa G, Castelnuovo G, Reid C. Videotherapy and therapeutic alliance in the age of COVID-19. Clin Psychol Psychother 2021;28(2):409-21.

(23) Tseng E, Freed D, Engel K, Ristenpart T, Dell N. A digital safety dilemma: Analysis of computer-mediated computer security interventions for intimate partner violence during COVID-19. In: Proceedings of the 2021 CHI conference on human factors in computing systems. New York: ACM, 2021.

(24) Slakoff DC, Aujla W, PenzeyMoog E. The role of service providers, technology, and mass media when home isn't safe for intimate partner violence victims: Best practices and recommendations in the era of COVID-19 and beyond. Arch Sex Behav 2020;49(8):2779–88.

(25) Nair VS, Banerjee D. "Crisis within the walls": Rise of intimate partner violence during the pandemic, indian perspectives. Front Glob Women's Health 2021;2:1–7.

(26) Montesanti S, Ghidei W, Silverstone P, Wells L. Examining the use of virtual care interventions to provide trauma-focused treatment to domestic violence and sexual assault populations. Ottawa, ON: CIHR 2020:1-40.

(27) Khanlou N, Ssawe A, Vazquez LM, Pashang S, Connolly JA, Bohr Y, et al. COVID-19 pandemic guidelines for mental health support of racialized women at risk of gender-based violence: Initial knowledge synthesis report. Ottawa, ON: CIHR 2020:1-36.

(28) LaPlante LM. Intimate partner violence: Assessment in the era of telehealth. Curr Psychiatry 2021;20(10):39–40.

(29) Simon MA. Responding to intimate partner violence during telehealth clinical encounters. JAMA 2021;325(22):2307-8.
(30) Evans ML, Lindauer M, Farrell ME. A pandemic within a pandemic – intimate partner violence during Covid-19. N Engl J Med 2020;383(24):2302-4.
(31) USPSTF. Intimate partner violence, elder abuse, and abuse of vulnerable adults: Screening. US Preventive Services Taskforce, 2018. URL: https://www.uspreventiveservicestaskforce.org/uspstf/recommendation/intimate-partner-violence-and-abuse-of-elderly-and-vulnerable-adults-screening.
(32) Dicola D, Spaar E. Intimate partner violence. Am Fam Physician 2016;94(8):646–51.
(33) Taylor A. Telehealth, Covid-19, and social determinants of health: Trauma informed care and intimate partner violence. California Telehealth Resource Center, 2021. URL: https://caltrc.org/news/telehealth-covid-19-and-social-determinants-of-health-trauma-informed-care-and-intimate-partner-violence/.

Chapter 4

Mental health stigma and psychiatric services utilization: A narrative review of impacting factors

Ajay Tom Eapen, MD, MPH
Sara Elzibak, BSc
and Satesh Bidaisee*, DVM, MSPH, EdD
Public Health and Preventive Medicine, St George's University, Grenada, West Indies

Abstract

About 47% of the population in the United States is affected by mental health disorders. There are many factors that play into the stigma associated with mental health. This chapter will explore the stigma associated with mental health disorders along with its associated factors and how they impact the utilization of mental health services. We conducted a narrative review synthesizing the findings from a literature search of medical databases including Google Scholar, St George's University Discovery Service, PubMed and the Cochrane Library. Keywords searched include "mental health stigma," "psychiatric disorders stigma" and a combination of those words. The studies were manually reviewed and categorized based on the explored factors associated with mental health stigma. The incidence of mental health disorders has increased over the last decades, yet multiple barriers prevent access to treatment. The attitudes associated with mental health stigma influences several aspects of the daily life of a mentally ill person:

* ***Correspondence:*** Satesh Bidaisee, Professor, Public Health and Preventive Medicine, St George's University, Grenada, West Indies. Email: sbidaisee@sgu.edu

In: Public Health: Intersection of Health, Humans, Animals and the Environment
Editors: Satesh Bidaisee and Joav Merrick
ISBN: 979-8-89530-443-3

self-perception, finances and housing, social and professional relationships, and treatment-seeking behavior. The stigma of mental illness must be considered holistically to truly understand the impact on those mentally ill. The negative consequences of stigma, although not completely revealed, are pervasive enough to negatively impact those with mental health issues.

Introduction

Mental health disorders affect nearly 47% of the population in the United States (1). Of the spectrum of mental health disorders nearly 22% are considered serious, 37% moderate, and 47% are considered mild in severity (2). Anxiety makes up 28.8% of these disorders followed by disruptive behaviors at 24.8%, mood disorders 20.8%, depression 16.6%, substance abuse 14.6%, alcohol abuse 13.2%, specific phobias 12.5% and social phobias at 12.1% (2). There are many reasons why individuals do not seek treatment including being disabled by the condition that they are unable to seek care or not being able to self-identify symptoms and seek treatment (3-5). Patients might also delay or avoid treatment due to fears of the stigma behind mental health disorders (6). The stigma associated with mental health disorders is a key barrier to patients seeking and completing treatment (7, 8). Patients will often feel discriminated or prejudiced against due to their mental illness (1, 9).

Stigma occurs when an individual is being viewed in a negative way due to a defining or distinguishing characteristic or personal trait (10). Canadian sociologist Erving Goffman (1922-1982) defined stigma as "an attribute that is deeply discrediting" to the individual that reduces someone "from a whole and usual person to a tainted, discounted one" (7). Goffman further defined six dimensions of stigma: concealability, course, disruptiveness, peril, origin, and aesthetics (7, 11). There are also three different levels of stigma: social-stigma, self-stigma, and healthcare professional-stigma (12, 13). Stigma arises from a combination of these factors either independently or simultaneously (11).

The purpose of this study is to present a narrative review of the available literature on the stigma associated with mental health disorders and factors impacting the utilization of psychiatric services.

Our review

The initial research question was formulated by planning on exploring mental health services. After reviewing very broadly on current issues associated with mental health services, the problem statement was refined. Stigma associated with mental health disorders has been widely known yet the specific factors impacting the utilization of psychiatric services has not been thoroughly explored.

For this narrative review, the following databases were used: Google Scholar, St George's University Discovery Service, PubMed, and the Cochrane Library. Keywords used to search include "mental health stigma," "psychiatric disorders stigma," "stigma mental illness" and a combination of those words individually and jointly. Literature was chosen from publications with empirical research ranging from January 2000 to January 2022. The search parameters were broad enough to make a comprehensive assessment of all relevant studies but narrow enough to focus on the proposed research question.

Inclusion criteria included papers with any analysis of factors associated with stigma, definitions of stigma, and exploration of mental health stigma. Studies included were not limited to the United States.

Papers were excluded if published before the year 2000 and studies that did not elaborate on any factors leading to mental health stigma or provide any statistics on mental health services utilization.

The studies were manually reviewed and categorized based on the explored factors associated with mental health stigma. A comprehensive overview of the literature was completed, and the information was placed into perspective for this study.

Discussion

The incidence of mental health disorders has increased over the last decades, yet multiple barriers prevent access to treatment. Many policies arising from state legislatures continue to cut funding to mental health and substance abuse treatment even though the demand is growing (2, 7). In addition, access to treatment is further limited by geographical location; many providers are concentrated in highly populous, opulent urban areas. This implies that those in rural areas are less likely to obtain proper treatment (14, 15). In regard to racial identities, those considered ethnic minorities have lower rates of treatment compared to European Americans (16). Ethnic minorities may also experience issues with differences in languages, underrepresentation of mental health professionals of color, and past experiences of racism (17, 18). The severity of illness also impacts the utilization of mental health services (19, 20). Systemic barriers would include finances, insurance coverage and transportation issues. All these factors may combine resulting in a lower utilization of mental health services.

Stereotypes and stigma are interrelated on multiple levels. Although theoretically varied, stereotypes constitute the public's perceptions of those mentally ill and contribute to the individual's stigma. Discriminatory or prejudicial attitudes add to the stigmatized status mentally ill patients experience and is a significant barrier to treatment (14, 15). In general, the public constructs a stereotyped or prejudicial image of a mentally ill person based on public perception of certain cues to an individual's mental status (7).

The attitudes associated with mental health stigma influences several aspects of the daily life of a mentally ill person: self-perception, finances and housing, social and professional relationships, and treatment-seeking behavior. Mental health stigma affects an individual's capacity to meet basic life aspects which results in an impact on the individual's desire and ability to seek mental health services (21, 22). In addition, there may be an interplay with self-perception which is also a mediator in the treatment-seeking sphere. In whole, the factors associated with stigma decreases an individual's willingness and ability to seek treatment. Below the factors impacting the utilization of mental health services is explored.

Mental health stigma decreases both self-efficacy and self-esteem (2, 7, 23). In patients with schizophrenia or affective disorders, increased stigma leads to lower self-efficacy and self-esteem (15). These results confirm previous studies exploring stigma's impact on self-perception (2, 7). Self-perception is therefore an intermediary between mental health stigma and seeking and/or maintaining treatment.

Individuals with mental illnesses often have financial and housing issues (3). Once an individual internalizes stigma, the impact on occupation is noteworthy. Several studies have shown interference with employment searching and the ability to perform assigned workplace duties (18 ,21). Those with mental illnesses are more likely to be unemployed or underemployed than the general population (14). 48-73% of individuals with a mental illness are employed compared to 76-87% of those who do not have a mental illness (2). Employees report increased incidences of workplace discrimination due to their health status (12). Considering the mediatory role of self-perception, being unemployed can lead to decreased drive to seek treatment. Additionally, lack of employment limits an individual's access to affordable healthcare and therefore the ability to receive adequate treatment. When seeking housing, mentally ill individuals may report experiences of racism and prejudice due to their health status (7, 16). Over 32% of those mentally ill experienced discrimination while seeking housing (2). These factors in whole contribute to those mentally ill either avoiding or delaying treatment for their respective disorder.

The stigma associated with mental health negatively affects interpersonal relationships with family and friends (24). Studies have shown that in bipolar or schizoaffective patients, those who perceived a higher level of stigma experienced impaired social functioning (10, 15). Mental health stigma is not limited to close relationships but has also been demonstrated in mental health practitioners (25). For example, social workers who perceived those mentally ill negatively were found to desire more social distance (2, 7). These stigmatizing attitudes are also present in nurses, students, doctors, and pharmacists (2). Although not all mental health professionals stigmatize, there are dire consequences for the patient when professionals do. The social network and quality of life of those mentally ill is impacted by experiences of discrimination or prejudice. Mentally ill individuals need a strong social support system to tackle their health issue, yet the stigma associated with mental health services may erode or prevent that structure from forming. This can lead to either avoidance or delay seeking treatment.

Conclusion

There has been much research into the relationship of mental health stigma and the individual's ability to receive adequate treatment. All studies explored highlighted the impact of mental health stigma on social and personal relationships. The factors impacting mental health stigma have not been clearly defined. There is an interplay between many factors, each uniquely connected resulting in adverse effects for those mentally ill. Most of the research is theoretical; a need for a detailed model of mental health stigma is crucial to understand the impact on those mentally ill.

The goal of this paper was to explore the impacting factors of mental health stigma on an individual seeking treatment. This study is a narrative review and is therefore limited in its scope. Mental health stigma is part of a complex series of factors including self-image, employment and housing, and personal and professional relationships. When considered in conjunction, these factors negatively impact those mentally ill. Around one in three adults in the United States will experience a mental health issue (7). Mental disorders result in a high rate of comorbidity so treatment must be available and accessible. There is a pronounced need for mental health services, yet many barriers hinder treatment-seeking behavior.

The stigma of mental health must be considered holistically to truly understand the impact on those mentally ill. Most studies have focused on those with severe mental illness as the symptoms and impacts are more apparent. Research should be done on those in their youth, of varying ethnic backgrounds, and with different degrees of illness severity. Very few studies examined the impact of specific ethnic and cultural attitudes toward mental health services and the consequences for the mentally ill individual. In addition, several studies vary in their approach to elucidating treatment seeking; some studies simply assessed attitudes while others recorded behaviors.

Overall, the treatment seeking relationship needs to be explored in a more comprehensive manner. Empirical testing and exploration of causal relationships needs to be undertaken. There may be several undiscovered factors, internal and external, that play a role in the relationship between treatment seeking and mental health stigma. More research needs to be completed to develop a more causal and thorough model. The negative consequences of stigma, although not completely revealed, are pervasive enough to negatively impact those with mental health issues. A mentally ill individual's social, psychological, and physical functioning is impacted by stigma resulting in a significant barrier to treatment. Much of the burden of mental illness on society and the individual can be ameliorated with future research and interventions.

References

(1) Rowan AB, Grove J, Solfelt L, Magnante A. Reducing the impacts of mental health stigma through integrated primary care: An examination of the evidence. J Clin Psychol Med Settings 2021;28(4):679–93.

(2) Sickel AE, Seacat JD, Nabors NA. Mental health stigma update: A review of consequences. Advances Ment Health 2014;12(3): 202–15.

(3) Corrigan PW. Mental health stigma as social attribution: Implications for research methods and attitude change. Clin Psychol Sci Pract 2000;7(1):48–67.

(4) DeLuca J, Akouri-Shan L, Jay SY, Redman SL, Petti E, Lucksted A, et al. Predictors of internalized mental health stigma in a help-seeking sample of adolescents and young adults experiencing early psychosis: The roles of psychosis-spectrum symptoms and family functioning. J Abnorm Psychol 2021;130(6):587-93.

(5) Bharadwaj P, Pai MM, Suziedelyte A, et al. Mental health stigma. Econ Lett 2017;159:57-60.

(6) Fox AB, Earnshaw VA, Taverna EC, Vogt D. Conceptualizing and measuring mental illness stigma: The mental illness stigma framework and critical review of measures. Stigma Health 2018;3(4):348-76.

(7) Ahmedani BK. Mental health stigma: Society, individuals, and the profession. J Soc Work Values Ethics 2011;8(2):41-416.

(8) Vidourek RA, Burbage M. Positive mental health and mental health stigma: A qualitative study assessing student attitudes. Ment Health Prev 2019;13:1-6.

(9) Mak WWS, Poon CYM, Pun LYK, Cheung SF. Meta-analysis of stigma and mental health. Soc Sci Med 2007;65(2):245-61.

(10) Smith RA, Applegate A. Mental health stigma and communication and their intersections with education. Commun Educ 2018;67(3):382-93.

(11) Shim R, Rust G. Primary care, behavioral health, and public health: Partners in reducing mental health stigma. Am J Public Health 2013;103(5):774-6.

(12) Mehta N, Clement S, Marcus E, Stona AC, Bezborodovs N, Evans-Lacko S, et al. Evidence for effective interventions to reduce mental health-related stigma and discrimination in the medium and long term: Systematic review. Br J Psychiatry 2015;207(5):377-84.

(13) Turner RN, Wildschut T, Sedikides C, Gheorghiu M. Combating the mental health stigma with nostalgia. Eur J Social Psychol 2013;43(5):413-22.

(14) Picco L, Abdin E, Pang S, Vaingankar JA, Jeyagurunathan A, Chong SA, et al. Association between recognition and help-seeking preferences and stigma towards people with mental illness. Epidemiol Psychiatr Sci 2018;27(1):84–93.

(15) Reynders A, Kerkhof AJ, Molenberghs G, Van Audenhove C. Attitudes and stigma in relation to help-seeking intentions for psychological problems in low and high suicide rate regions. Soc Psychiatry Psychiatr Epidemiol 2014;49(2):231-9.

(16) Boxell O. Social context affects mental health stigma. Open Health 2020;1(1):29-36.

(17) Goepfert NC, Conrad Von Heydendorff S, Dreßing H, Bailer J. Effects of stigmatizing media coverage on stigma measures, self-esteem, and affectivity in persons with depression – An experimental controlled trial. BMC Psychiatry 2019;19(1):138.

(18) Goh YS, Ow Yong QYJ, Tam WW. Effects of online stigma-reduction programme for people experiencing mental health conditions: A systematic review and meta-analysis. Int J Ment Health Nurs 2021;30(5):1040-56.

(19) Schnyder N, Panczak R, Groth N, Schultze-Lutter F. Association between mental health-related stigma and active help-seeking: Systematic review and meta-analysis. Br J Psychiatry 2017;210(4):261-8.

(20) Stolzenburg S, Freitag S, Evans-Lacko S, Speerforck S, Schmidt S, Schomerus G. Individuals with currently untreated mental illness: Causal beliefs and readiness to seek help. Epidemiol Psychiatr Sci 2019;28(4):446-57.
(21) Horsfield P, Stolzenburg S, Hahm S, Tomczyk S, Muehlan H, Schmidt S, et al. Self-labeling as having a mental or physical illness: The effects of stigma and implications for help-seeking. Soc Psychiatry Psychiatr Epidemiol 2020;55(7):907-16.
(22) Schomerus G, Stolzenburg S, Freitag S, Speerforck S, Janowitz D, Evans-Lacko S, et al. Stigma as a barrier to recognizing personal mental illness and seeking help: A prospective study among untreated persons with mental illness. Eur Arch Psychiatry Clin Neurosci 2019;269(4):469-79.
(23) Clement S, Schauman O, Graham T, Maggioni F, Evans-Lacko S, Bezborodovs N, et al. What is the impact of mental health-related stigma on help-seeking? A systematic review of quantitative and qualitative studies. Psychol Med 2015;45(1):11-27.
(24) Arboleda-Flórez J, Stuart H. From sin to science: Fighting the stigmatization of mental illnesses. Can J Psychiatry 2012;57(8):457-63.
(25) Wei Y, McGrath PJ, Hayden J, Kutcher S. Mental health literacy measures evaluating knowledge, attitudes and help-seeking: A scoping review. BMC Psychiatry 2015;15:291.

Chapter 5

Barriers to preventing premature mortality in rheumatoid arthritis among Hispanics with low socioeconomic status in the United States: A narrative review

Melissa C Andrade, MD, MPH
Sara Elzibak, BSc
and Satesh Bidaisee*, DVM, MSPH, EdD
Public Health and Preventive Medicine, St George's University, Grenada, West Indies

Abstract

Data from the last 15 years have demonstrated that when autoimmune diseases are considered as a group, they rank among the 10 leading causes of death among women under age 75 years. Rheumatoid arthritis (RA) is a chronic autoimmune arthritic disease associated with progressive joint damage, diminished quality of life, disability, and premature mortality. Our review was conducted from June 2020 to February 2021. A total of 108 articles were generated from the initial search with 21 full articles were used to complete the review. Three barriers to the prevention of premature mortality in United States Hispanics were identified: low-income, low education attainment, lack of access to specialized care and early start of treatment. Results show that the low average income of Hispanics is an obstacle to receiving timely and appropriate health care. Low-income people are less able to afford the out-of-pocket costs of care, even if they have health insurance coverage. Low education may impair

* *Correspondence:* Satesh Bidaisee, Professor, Public Health and Preventive Medicine, St George's University, Grenada, West Indies. Email: sbidaisee@sgu.edu

In: Public Health: Intersection of Health, Humans, Animals and the Environment
Editors: Satesh Bidaisee and Joav Merrick
ISBN: 979-8-89530-443-3

people's ability to navigate the complex health care delivery system, communicate with health care providers, and understand providers' instructions. Lack of insurance also contributes to many Hispanic not having a usual source of care such as a primary care physician and thus makes early referral to specialized care, specifically a rheumatologist, much more difficult. Lastly, the number of practicing rheumatologists does not meet the current or future needs of the patient population, patients are waiting longer for appointments.

Introduction

Rheumatoid arthritis (RA) often strikes small joints in the wrists, hands and feet. Sometimes RA can affect elbows and knees or organs, such as the eyes, skin or lungs. About 75% of RA patients are women (1). It usually starts between ages 30 and 60 but can happen to people at any age. By conservative estimates, between 2010-2012 about four million Hispanic adults in the United States had doctor-diagnosed arthritis (2). Arthritis is the most prevalent chronic condition in women and individuals with family incomes of less than $20,000, and the leading cause of activity restriction among women (3). In 2019, the United States Census Bureau reported that 10.7% of Hispanics in the United States had an annual income of less than $15,000 (4). Lack of improved management of RA and prevention of comorbidities associated with RA also cause a significant burden on the United States labor force. A 2010 study found that about one-fourth to one-half of all patients with RA become unable to work within 10 to 20 years of follow-up after diagnosis (2). Not only is RA a chronic, life-long, and costly illness, it is of major public health concern because RA patients have been shown to be at increased risk of other important comorbidities such as cardiovascular disease, pulmonary fibrosis, and lymphoma (5). A 2007 study found that excess mortality in RA has been seen in cardiovascular disease (31 percent), pulmonary fibrosis (four percent), and lymphoma (2.3 percent) (2). Recent research has called attention to the unique clinical features of RA and self-reported outcomes in Hispanic minorities, especially in low socioeconomic status, uninsured, and immigrant populations collectively described as "vulnerable patients." The rate of activity-limitation attributable to arthritis is higher among Hispanic patients. This likely reflects the poorer socioeconomic conditions and lack of health insurance that prevail among Hispanic populations, which may limit their access to rheumatologic care (6). There are many different variables implicated in why United States Hispanics with RA die prematurely; the purpose of this narrative review is to explore if variables such as low income, low education attainment, and lack of access to specialized care/early start of treatment are significant barriers to proper care and contributing to premature mortality in those diagnosed with RA.

Our review

This narrative review was conducted to answer the following question: What are barriers encountered by low socioeconomic status Hispanics in the United States diagnosed with rheumatoid arthritis to preventative care and avoidance of premature mortality?

The review was conducted from June 2020 to February 2021. Articles searched in Google Scholar general search engine and PubMed database were used; using Medical Subject Headings (MeSH) keywords: rheumatoid arthritis (RA), socioeconomic status (lack of health insurance, Medicaid, low SES, low income), clinical outcomes (premature mortality, functional disability), treatment delay (Disease-Modifying Antirheumatic Drugs, DMARD, DMARD lag), burden (cost, united state healthcare, united states labor force), and Hispanic (United States).

Process of literature selection was reported following the PRISMA (Preferred Reporting Items for Systemic Reviews and Meta-analysis) statement guidelines (see Figure 1). Analysis was done according to ethnicity status of Hispanic defined as "persons whose ancestry originated from one of the Spanish-speaking countries" (6). The term Latino is a descriptor preferred by some authorities, considered synonymous with Hispanic. The term Hispanic is used because it is the most frequently used in the health care literature. Those with low socioeconomic status defined as annual income less than $20,000, lack of health insurance or covered by state-funded insurance programs such as Medicare were then analyzed. Finally, clinical outcomes such as premature death, delay in receipt of disease-modifying antirheumatic drugs, and increased disability (activity-limitation) were analyzed. A total of 108 articles were generated from the initial search. 50 non-duplicate articles were then screened via title and abstract. After inclusion and exclusion was applied, 25 articles were excluded. 25 articles were retrieved, and their full text was screened. After inclusion and exclusion was applied, two articles were excluded, and during data extraction, two articles were excluded. Ultimately, 21 full articles were used to complete this narrative review.

Articles were selected using the following inclusion criteria: publication between 2000 and 2020, available as full text in English, all types of articles, exclusively related to humans, all genders, all ages, Hispanic patients, and dealing exclusively with the diagnosis of rheumatoid arthritis.

Articles were excluded if all inclusion criteria were not met, there was no mention of Hispanic patients, and focus on diseases other than rheumatoid arthritis.

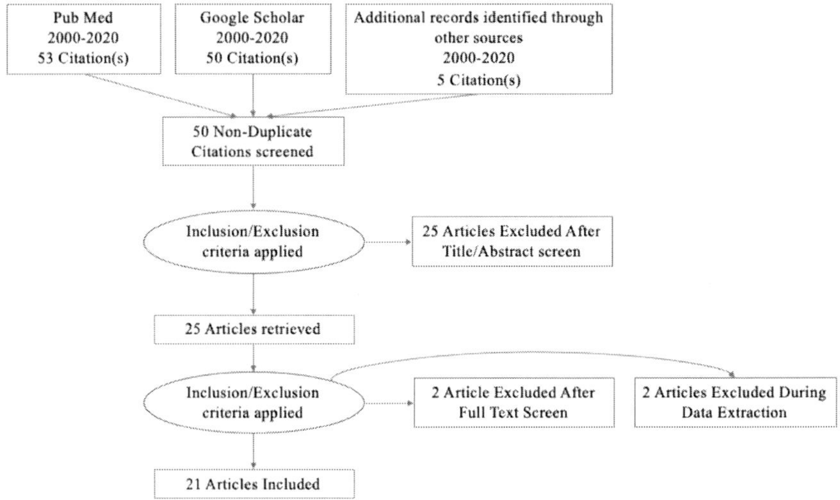

Figure 1. Literature screening led to the evaluation of 21 full-length articles following the PRISMA (Preferred reporting items for systemic reviews and meta-analysis) statement guidelines.

Qualitative findings were organized into Table 1, Barriers to prevention of premature mortality in rheumatoid arthritis in low socioeconomic Hispanics in the United States. Three barriers were used: low-income, low education attainment, lack of access to specialized care and early start of treatment. Then using data extracted from this review, the effect and result that these barriers have on prevention of premature mortality in rheumatoid arthritis is summarized. Finally, recommendations on how to combat these barriers and better manage rheumatoid arthritis in low socioeconomic Hispanics are stated.

Table 1. Barriers to prevention of premature mortality in rheumatoid arthritis in low socioeconomic Hispanics in the United States

Barriers to prevention of premature mortality in RA	Effect mechanism	Results	Number of Articles	Recommendations
Low-income limits access to timely and appropriate care of rheumatoid arthritis.	Inability to pay for medications and specialized care	The low average income of Hispanics is an obstacle to receiving timely and appropriate health care (7). Low-income people are less able to afford the out-of-pocket costs of care, even if they have health insurance coverage (7).	11	Understand the barriers to health care in Hispanics with low income. Promote access to a prompt diagnosis and treatment, developing algorithms according the realities of those with low-income and lack of health insurance.
Low education attainment	Lack of awareness of comorbidities associated with rheumatoid arthritis and their implications on survival	Low education may impair people's ability to navigate the complex health care delivery system, communicate with health care providers, and understand providers' instructions (7).	5	Educate patients and doctors on the comorbidities contributing to premature death in those diagnosed with RA.
Lack of access to specialized care, specifically access to a rheumatologist.	Limited rheumatologists, long follow-up times and delay in start of disease-modifying antirheumatic drugs (DMARD) therapy	Lack of insurance also contributes to many Hispanic not having a usual source of care such as a primary care physician and thus makes early referral to specialized care, specifically a rheumatologist, much more difficult (7). The number of practicing rheumatologists does not meet the current or future needs of the patient population, patients are waiting longer for appointments (8).	6	Harmonize the treatment between providers and RA Hispanic patients in the United States by developing a program in community health centers or health fairs focused on RA education and assistance in obtaining appropriate care.

Findings

There were three barriers to prevention of premature mortality in RA among low socioeconomic status Hispanics in the United States identified in this narrative review. The first barrier was low-income. Results showed that the low average income of Hispanics is indeed an obstacle to receiving timely and appropriate health care (7). In addition, low-income people are less able to afford the out-of-pocket costs of care, even if they have health insurance coverage (7). The second barrier identified was low educational status among Hispanics in the United States. The results showed that this may impair people's ability to navigate the complex health care delivery system, communicate with health care providers, and understand providers' instructions (7). The third barrier identified was lack of access to specialized care and early start of treatment. Results showed that lack of insurance contributes to many Hispanic not having a usual source of care such as a primary care physician and thus makes early referral to specialized care, specifically a rheumatologist, much more difficult (7). In addition to this, the current number of practicing rheumatologists does not meet the current or future needs of the patient population, patients are waiting longer for appointments (8).

Discussion

Generally, results showed that low income and commonly associated social factors such as lack of educational attainment, and delay start of DMARD therapy in United States Hispanics is a barrier to prevention of premature death in those diagnosed with rheumatoid arthritis.

The low average income of Hispanics is an obstacle to receiving timely and appropriate health care. Low-income people are less able to afford the out-of-pocket costs of care, even if they have health insurance coverage (7), and many low-income people lack health insurance altogether. In 2019 the United States Census Bureau reported that 10.7% of Hispanics in the United States had an annual income of less than $15,000 (4). Lifetime per patient cost of those with RA is estimated to range between $61,000 and $122,000. Indirect costs are over twice those of direct expenses, leading to total RA-related costs of approximately $26 to $32 billion annually in the United States alone (9). Part of the reason that RA is such a costly disease for patients to treat is the need for medications to control the disease early on. Premature death has long been recognized as a manifestation of RA and several studies have shown that early treatment is critical for optimal care in these patients (10). Several studies describing the occurrence of irreversible joint damage in early disease supported the hypothesis that the earlier use of disease-modifying antirheumatic drugs (DMARD) might lead to earlier disease control and reduced joint damage (11). In one nonrandomized comparison, early introduction of DMARD was associated with a better disease outcome after 2 years (11). Multiple guidelines recommend initiation of DMARD treatment as early as possible after symptom onset, ideally during the initial 4 to 5 months when the disease is most susceptible to treatment. However, despite the evidence and recommendations, the median time for treatment initiation remains suboptimal and unsatisfactory, especially in minorities and lower-income patients (12). Recent population-based studies of DMARD use in patients with RA report consistently low rates of DMARD receipt (30%-52%) (5). Research has shown that on average, patients with lower SES waited 8.5 ± 10.2 years after onset of RA symptoms to begin DMARD treatment, compared to those in middle and upper SES tertials who waited 6.1 ± 7.9 years (P=0.002) and 6.1 ± 8.6 years (P=0.009), respectively (10). Hispanics face a variety of barriers both financial and nonfinancial when it comes to receiving adequate healthcare in a timely manner. Degree of acculturation, language, and immigration status all directly affect access to care. Recent arrivals to the United States are more likely to be isolated from mainstream U.S. society and to be unfamiliar with the U.S. health care system, a situation that may interfere with obtaining appropriate and timely care (7). The need for appropriate care in those diagnosed with RA is further highlighted by the fact that lack of appropriate care leads to rapid disability and often, the complete loss of income altogether. Loss of income leads to the disease being an even greater financial burden on these low-income families and because of its progressive nature, many individuals report missing work, losing substantial amounts of productivity or choose not to work because of disease-related disabilities (2). Approximately 20 to 70 percent of individuals who were working at the inception of their RA were disabled after seven to 10 years (2). Total national arthritis-attributable lost wages were $164 billion in 2013. That's $4,040 less pay for an adult with arthritis compared with an adult without arthritis. The high earnings losses were because of the substantially lower percentage of adults with arthritis working compared with adults without arthritis. This indicates the need for interventions that keep people with arthritis in the work force (13).

As the United States ethnically diverse population continues to grow, poor health outcomes in minorities represent a major public health problem. Marked differences in disease prevalence, manifestation, burden, management, and treatment between Caucasian and minority groups have been well documented. The less favorable health status of minorities has been attributed, in part, to factors such as lower levels of education and income (14). In the United States, educational attainment and income are the indicators that are most commonly used to measure the effect of socioeconomic position on health. Research indicates that substantial educational and income disparities exist across many measures of health (15). According to the Centers for Disease Control, in the 2011 population aged ≥25 years showed statistically significant absolute disparities in noncompletion of high school (16). The absolute racial/ethnic difference between non-Hispanic whites and each of the other racial ethnic groups was highest for Hispanics (30.4 percentage points). Noncompletion of high school increased with increasing poverty; the absolute difference for the poorest group was approximately three times the absolute difference for the middle-income group (6.4 versus 1.7 percentage points) (15). Low education may impair people's ability to navigate the complex health care delivery system, communicate with health care providers, and understand providers' instructions (7). Studies have examined the effects of low socioeconomic status, including low level of education, on the progression of RA. Low socioeconomic status is associated with higher RA susceptibility, worse outcomes and less use of health services (17). Patients with lower socioeconomic status also have longer delays in DMARD initiation (i.e., 8.5 years for low socioeconomic status vs six years for the middle and upper groups (12). Low educational attainment among low-income Hispanics emphasizes the need to educate patient on the complexities of RA, its sequalae, and the need for preventative care to avoid premature death. Several comorbidities have been identified which could explain why the mortality gap between RA and the general population appears to be widening (18). These conditions include cardiovascular disease (CVD), infection, malignancy, gastrointestinal (GI) disease, and osteoporosis leading to fracture (9). Not only do patients with RA have a higher risk of multiple comorbid conditions but also they tend to experience worse outcomes after the occurrence of these comorbid illnesses, in part because of the systemic inflammation and immune dysfunction associated with RA that appears to promote and accelerate comorbidity and mortality (18). These comorbidities demonstrate the importance of educating patients with RA, especially those with low educational attainment who may not be aware of the comorbidities that may accompany their RA diagnosis, on the increased need for optimization of primary and secondary preventative health visits and screening for comorbidities.

Lack of health insurance is a financial barrier for many low-income Hispanics and contributes to the delay in seeing a rheumatologist and treatment initiation of DMARD therapy. Health insurance reduces the out-of-pocket costs of health care and has been shown to be the single most important predictor of utilization. Without health insurance coverage, many people find health care unaffordable and forgo care even when they think they need it (7). Historically, lack of health insurance coverage has been a major problem for Hispanics, who are substantially more likely to be uninsured than non-Hispanic whites. For example, in 2004, 36 percent of Hispanics under age 65 lacked health insurance coverage, compared with 15 percent of whites (7). Reasons for the low health insurance coverage among Hispanics are many, but research does show that Hispanic males in poor, low-income, or middle-income households; those earning low wages; and those in firms with fewer than 25 workers were less likely than their white peers to have employer-provided insurance (7). Early referral of patients with RA to a rheumatologist is a key part of optimal clinical care, and recommendation guidelines have been issued in the US and Canada because prompt use of DMARDs can prevent joint destruction (19). Lack of insurance also contributes to many Hispanic's continuity of care with a primary care physician and thus makes early referral to specialized care, specifically a rheumatologist, much more difficult (7). Despite the demonstrated benefits, many patients do not receive proper treatment early enough to benefit from rheumatologic care (19). In 2007, a study was conducted that examines disparities in disability, pain, and global health between Caucasian, African Americans, and Hispanic patients with rheumatoid arthritis. Their results showed the number of comorbid conditions and DMARD were similar across groups, but there was slightly higher use of DMARD by Caucasian patients with RA compared to both African Americans and Hispanics ($p < 0.01$ and $p < 0.05$). In sum, Hispanics were the youngest, least educated, and had developed their RA earlier than Caucasians or African Americans. Global health scores were worse for both ethnic groups compared to Caucasians, but only statistically different for Hispanics ($p < 0.05$). For all 3 measures, Hispanics had the worst scores (14). Their results highlight the need to emphasize education of their condition to Hispanic populations.

In addition to low income and lack of insurance, results of the first rheumatology workforce study in the past 10 years were released last

November, confirming concerns that the supply of rheumatologists may not meet demand in the near future. The extensive report, commissioned by the ACR, lays out the hard facts about this medical specialty that its practitioners had long anticipated were true—the number of practicing rheumatologists does not meet the current or future needs of the patient population, patients are waiting longer for appointments, and practices must be redesigned to preserve and improve the quality of the profession (8). The ACR projected that the number of rheumatologists in adult practice will increase only 1.2% between 2005 and 2025. However, the demand for rheumatologists to treat patients in that time period will rise 46% (8). Another study used data from 152 DMARD naïve RA patients; 35% were white, 37% black, 20% Hispanic, and 8% other. The range in median time to first rheumatology visit was 6 to 8 months for all patient groups, except Hispanic. This group had a median time of 22.7 months ($p = 0.01$) (12). There is significant delay in initial presentation to a rheumatologist that was associated with a higher disease severity at presentation, especially for Hispanic patients (12).

Conclusion

The purpose of this narrative review was to identify and better understand barriers to the prevention of premature mortality in low socioeconomic status Hispanics in the United States diagnosed with rheumatoid arthritis. The key barriers identified in this narrative review were low income, low education attainment, and lack of access to specialized care and early start of treatment.

It is of financial and public health concern to have more health promotion and prevention of premature mortality in low socioeconomic status Hispanics diagnosed with RA in the United States because in the end it could lessen the financial burden of arthritis on the United States healthcare system and prevent premature death among Hispanics diagnosed with RA. Lifetime per patient cost of those with RA is estimated to range between $61,000 and $122,000. Indirect costs are over twice those of direct expenses, leading to total RA-related costs of approximately $26 to $32 billion annually in the United States alone (9). Rheumatoid arthritis also proves to be a burden on the United States workforce. Approximately 20 to 70 percent of individuals who were working at the inception of their RA were disabled after seven to 10 years (2). The Arthritis Foundation also reported a 2010 study that found that about one-fourth to one-half of all patients with RA become unable to work within 10 to 20 years of follow-up after diagnosis. The indirect cost of RA due to lost productivity has been estimated to be nearly three times greater than the costs of treating the disease (2).

Recommendations

Some recommendations to combat this lack of prevention of premature mortality in low socioeconomic status Hispanics in the United States are as follows. As a medical and professional community, we must firstly recognize and understand that there are barriers to health care in Hispanics with low income- specifically those diagnosed with RA. As a medical and public health community we must strive to promote access to a prompt diagnosis and treatment by developing algorithms according the realities of those with low-income and lack of health insurance. Third, education can go a long way especially in those Hispanics who recently immigrated or have a low educational status. Education of both patients and doctors on the comorbidities contributing to premature death in those diagnosed with RA is essential to their prevention. Fourth, recognizing that allowing a whole ethnic group of low income to progress to advanced rheumatologic disease does impact the United States healthcare economy. Providing these patients with more services can allow for earlier intervention and prevention of premature death and high costs. Finally, harmonizing the treatment between providers and RA Hispanic patients in the United States by developing a program in community health centers or health fairs focused on RA education and assistance in obtaining appropriate care is essential to preventing premature mortality.

References

(1) Marder W, Vinet É, Somers EC. Rheumatic autoimmune diseases in women and midlife health. Womens Midlife Health 2015;1(1):1-8.

(2) Arthritis Foundation. Arthritis by the numbers. Atlanta, GA: Arthritis Foundation, 2019.

(3) Abraido-Lanza AF. Latinas with arthritis: Effects of illness, role identity, and competence on psychological well-being. Am J Commun Psychol 1997;25(5):601-27.

(4) Semega J, Kollar M, Shrider EA, Creamer J; The United States Census Bureau. Income and poverty in the United States 2019. US Census Bureau, 2020. URL: https://www.census.gov/library/publications/2020/demo/p60-270.html.

(5) Schmajuk G, Trivedi AN, Solomon DH, Yelin E, Trupin L, Chakravarty EF, et al. Receipt of disease-modifying antirheumatic drugs among patients with rheumatoid arthritis in Medicare managed care plans. JAMA 2011;305(5):480-6.
(6) Escalante A, Del Rincón I. Epidemiology and impact of rheumatic disorders in the United States Hispanic population. Curr Opin Rheumatol 2001;13(2):104-10.
(7) Escarce JJ, Kapur K. Access to and quality of health care. In: Tienda M, Mitchell F, eds. Hispanics and the future of America. Washington, DC: National Academies Press, 2006:410-46.
(8) Hartnett T. Who will treat qrthritis in 2025? Rheumatologist 2007;1(1):12-3. URL: https://media.wiley.com/assets/1156/45/TRJanWebFinal.pdf.
(9) Mikuls TR, Saag KG. Comorbidity in rheumatoid arthritis. Rheum Dis Clin North Am 2001;27(2):283-303.
(10) Molina E, del Rincon I, Restrepo JF, Battafarano DF, Escalante A. Association of socioeconomic status with treatment delays, disease activity, joint damage, and disability in rheumatoid arthritis. Arthritis Care Res 2015;67(7):940-6.
(11) Lard LR, Visser H, Speyer I, Vander Horst-Bruinsma IE, Zwinderman AH, Breedveld FC, et al. Early versus delayed treatment in patients with recent-onset rheumatoid arthritis: Comparison of two cohorts who received different treatment strategies. Am J Med 2001;111(6):446-51.
(12) Riad M, Dunham DP, Chua JR, Shakoor N, Hassan S, Everakes S, et al. Health disparities among Hispanics with rheumatoid arthritis: Delay in presentation to rheumatologists contributes to later diagnosis and treatment. J Clin Rheumatol 2020;26(7):279-84.
(13) Centers for Disease Control and Prevention. The cost of arthritis in US Adults. Atlanta, GA: CDC, 2019. URL: https://www.cdc.gov/arthritis/data_statistics/cost.htm.
(14) Bruce B, Fries J, Murtagh K. Health status disparities in ethnic minority patients with rheumatoid arthritis: A cross-sectional study. J Rheumatol 2007;34(7):1475-9.
(15) Beckles GL, Truman BI. Education and income - United States, 2009 and 2011. MMWR Suppl 2013;62(3):9-19.
(16) Centers for Disease Control and Prevention. CDC health disparities and inequalities report - United States, 2013. MMWR 2013;62(Suppl 3):1-187.
(17) Jacobi CE, Mol GD, Boshuizen HC, Rupp I, Dinant HJ, Van Den Bos GA. Impact of socioeconomic status on the course of rheumatoid arthritis and on related use of health care services. Arthritis Rheum 2003;49(4):567-73.
(18) Gabriel SE. Why do people with rheumatoid arthritis still die prematurely? Ann Rheum Dis 2008;67(Suppl 3):30-5.
(19) Feldman DE, Bernatsky S, Haggerty J, Leffondré K, Tousignant P, Roy Y, et al. Delay in consultation with specialists for persons with suspected new-onset rheumatoid arthritis: A population-based study. Arthritis Rheum 2007;57(8):1419-25.
(20) Gabriel SE. The epidemiology of rheumatoid arthritis. Rheum Dis Clin North Am 2001;27(2):269-81.

(21) Karpouzas GA, Draper T, Moran R, Hernandez E, Nicassio P, Weisman MH, et al. Trends in functional disability and determinants of clinically meaningful change over time in Hispanic patients with rheumatoid arthritis in the US. Arthritis Care Res 2017;69(2):294-8.

Chapter 6

Restructuring urban zoning policies to reduce childhood obesity rates in New York City

Priya Mallikarjuna, MD, MPH
Sara Elzibak, BSc
and Satesh Bidaisee*, DVM, MSPH, EdD
Public Health and Preventive Medicine, St George's University, Grenada, West Indies

Abstract

The built environment has a significant impact on a child's health and future wellbeing. It is a long-held belief that disease prevention rests singularly on the individual and their motivation to adopt lifestyle changes. However, this is not true. Children are a prime example of how zoning contributes to unequal, inequitable, exclusionary neighborhood development and, subsequently, to stark health disparities amongst ethnic and socioeconomic groups. Zoning policies affect the makeup of communities in terms of green space, healthcare clinics and fresh food markets. The purpose of this capstone is to investigate and highlight the zoning policies throughout New York City and how such policies accentuate, or diminish, socioeconomic and healthcare disparities within the pediatric population in the five boroughs. Although zoning was historically created to protect the welfare of citizens, it has slowly evolved as an obstacle against equally distributed goods and services. New York City has thankfully moved towards positive changes in terms of healthy food access near schools and pushes for safer neighborhoods, however there is still a tremendous amount of progress to be made in

* *Correspondence:* Satesh Bidaisee, Professor, Public Health and Preventive Medicine, St George's University, Grenada, West Indies. Email: sbidaisee@sgu.edu

In: Public Health: Intersection of Health, Humans, Animals and the Environment
Editors: Satesh Bidaisee and Joav Merrick
ISBN: 979-8-89530-443-3

terms of food availability and safe, recreational spaces in low-income neighborhoods. The components addressed in this narrative review focus on these particular changes and how these can positively impact childhood health.

Keywords: environmental justice, zoning policy, New York City, healthcare disparities, poverty, access to healthcare, access to food, food deserts, pediatric health outcomes, childhood obesity, childhood metabolic syndrome, socioeconomic bracket, food swamps, fast food, BMI

Introduction

The built environment, in which one lives or grows up in, has a tremendous impact on health and wellbeing. Physicians and healthcare workers can counsel patients and families on healthy lifestyle changes and dietary needs; however, it is easy to overlook the problems of accessibility. Here, the conversation of zoning policy and urban planning arises. New York City created the first comprehensive zoning law in 1916 during the Industrial Era to separate land in hopes to limit harmful chemical exposures and thus improve public health (3). However, these policies evolved into a barrier against equally distributed goods/services and powered "exclusionary zoning" (1-3). Exclusionary zoning refers to the practice of separating minority or lower socioeconomic bracket residents to a particular area/zone. This has a host of downstream effects such as creating a greater density of fast-food restaurants, a lack of fresh produce markets (food deserts) and less green space/recreational areas (2, 4, 5).

As Wilson et al., (3) aptly put, "...the nation's obesity epidemic may be a result of the lack of neighborhood and metropolitan level infrastructure that supports physical activity, active lifestyles, and equity in healthy food access." This holds especially true for children. Not only are children heavily influenced by their surroundings, but they are also inherently one of the most vulnerable populations in our community. In the years 2017-2018, the prevalence of obesity in 2–19-year-olds was approximately 20%, with nearly a 10% difference comparing highest to lowest income brackets (6). Because poorer and minority communities are rich in "health restricting activities," it becomes difficult for residents and their families to lead healthy livelihoods (3). How can a child play safely in a neighborhood that is in close proximity to hazardous waste or sewer treatment plants? Or in a community where green space and recreational activities are limited? This built environment can either construct or deconstruct unhealthy lifestyle habits or positive health outcomes. The built environment is just that: built. By combining legislation, public health, and local businesses, we can effectively reduce negative influences on children's health and reduce risk factors for obesity.

The purpose of this review is to further elucidate the relationship between zoning policies in the five New York City boroughs and the downstream effects these policies have on childhood wellbeing and obesity. With this knowledge, the main objective is to re-imagine zoning and social policies to optimize pediatric health and reduce health inequity.

Methods

The narrative review was arranged into the following steps to obtain relevant data: First, the research question was identified to specify zoning policies in New York City (targeting the five boroughs of Manhattan, Staten Island, Queens, Bronx and Brooklyn) and its relationship with pediatric healthcare disparities and socioeconomic levels in these areas. Secondly, data was collected, via public search engines, of zoning policies in New York City and of communities with the highest/lowest incidence of metabolic syndromes, obesity, and levels of socioeconomic status. Thirdly, all relevant articles were sifted through based on the criteria elaborated upon below. Fourthly, the data was summarized/charted and subsequently reported.

Computer searches used the Google Scholar search engine to obtain articles on general zoning policies in New York City, along with an overview of social health determinants. Then, PubMed, Science Direct and Cochrane Library databases were used to obtain more specific articles pertaining to the following keywords: zoning policy, race, New York City, Manhattan, Queens, Brooklyn, Staten Island, Bronx, healthcare disparities, socioeconomic bracket, access to healthcare, fast food restaurants, metabolic disorders, childhood obesity and BMI. Relevant articles were selected from the years spanning from 2000 to 2022.

Papers were gathered from the initial Google Scholar general search and then assessed for relevance to the study hypothesis and quality of research demonstrated. Articles were included if the authors illustrated a relationship (either positive or negative) between either of the combinations listed: zoning policy and pediatric healthcare access, socioeconomic status and child/pediatric healthcare access, zoning and obesity/BMI, and lastly, zoning and prevalence of metabolic disorders in children. Articles were excluded if they describe research outside the boundaries of the New York City boroughs or described health outcomes outside the pediatric population. Special attention was made to include studies that delved into the deficits in current zoning policies in New York City and how these deficits could be realistically rectified. Five studies were chosen to highlight New York zoning policies and its effects on childhood obesity, as summarized in Table 1.

Table 1. Objectives of the five articles used to illustrate zoning policies in NYC and its relationship to childhood obesity

Article author(s) and year of publication	Objective
Schroeder, K. et al., 2020	Secondary analysis of longitudinal data looking at New York City public school students from kindergarten through 12th grade and their risk of obesity moving from higher to lower poverty neighborhood and vice versa (using BMI z score trajectories for obesity risk).
Neckerman, K. et al. (2010)	Examines food types in close proximity to schools in New York City, discrepancies between economic/race/ethnicity in access to healthy food options.
Elbel, B. et al. (2019)	Cross sectional study of 800,000 public school students, connecting food environment around homes to socioeconomic status and downstream health issues in New York City children.
Loureiro, M.I., Freudenberg, N. (2012)	Comparative case study looking at three different cities (Lisbon, London and New York City) with various health, political and zoning policies and how effective these policies are in reducing childhood obesity rates.
Rummo, P., et al. (2020)	Association of food outlets (particularly bodegas and corner stores) near New York City schools and likelihood outcome for obesity in high school students between 2009-2013 in these zoning areas.

Results

In Table 2, the results are reported for each study. A more in-depth conversation of deficits in the current models are described subsequently in the analysis section.

Table 2. Summary of current zoning policies in New York City and effects on childhood obesity

Article author(s) and year of publication	Current policies in New York City	Effects on childhood obesity
Schroeder, K. et al., 2020	Percent of individuals below Federal Poverty Level were geocoded and separated via ethnic background. Increased poverty entails more dangerous neighborhoods and lower density of healthy, yet cheap, food options	Females who moved from lower to higher poverty neighborhood had greater BMI *z*-score acceleration compared to those who moved from lower to higher poverty neighborhood. Similar results for males. Unhealthy conditions in conjunction with stress responses to unsafe neighborhoods can affect metabolic processes.
Neckerman, K. et al. (2010)	NYC public schools students have easy access to unhealthy foods near schools. Ethnic minority children were more likely than ethnic majority children to attend schools with unhealthy food options nearby. Small convenience stores/bodegas more prevalent near schools were lower socioeconomic status families and ethnic minority children live.	Although nearly all NYC public schools have energy dense, unhealthy food options available within 400m or 5 minutes walking distance, lower income and Hispanic students have a higher exposure to these food options amongst all groups studied.
Elbel, B. et al. (2019)	Neighborhood imbalances in food access, income key determinant of obesity along with what “food environment” children live in. Schools with higher percentage of ethnic minority groups have higher density of fast food restaurants in the area compared to schools with higher percentage of White students.	Black, Hispanic and Asian public school students live and go to school closer to corner stores, fast food restaurants etc., more so than White students regardless of socioeconomic status.
Loureiro, M.I., Freudenberg, N. (2012)	Importance of capacity building in low resource, low socioeconomic bracket locations and ethnic minority groups Focus on the poorer neighborhoods and target high retailing prices for healthy foods	Rates of overweight children are higher in Black ethnicity and in males (aged ~13 years) Higher caloric consumption, inadequate caloric expenditure due to unsafe neighborhoods
Rummo, P. et al. (2020)	Schools in disadvantaged areas/zones have higher density and availability of unhealthy food options Lower income students as well as Black and Hispanic students have greater access to these unfavorable food options regardless of food location, indicating the location of their household/neighborhood contains higher density of foods	Consumption of fast food linked to higher energy intake per meal and overall poorer diet (high in carb, high in sugar, low in nutritional value) in adolescents. Corner stores stock cheap, high calorie food that adolescents and school students purchase. Healthier options, if available, are more expensive and less favorable.

Discussion

It is important to acknowledge how the built environment contributes to healthcare disparities across socioeconomic and ethnic groups (3). The United States has a deep-seated history in residential and racial segregation where, nationally, there is a persisting divide and difference between where wealthier and predominantly Caucasian populations live and poorer, ethnic minorities live. New York City is hardly an exception. The neighborhood impact on health is multifactorial, depending on sociopolitical and cultural aspects that have downstream effects on childhood health and eventual obesity (7). As repeatedly showcased in the research articles above, minority neighborhoods are under-sourced in multiple areas, such as fresh food supermarkets and recreational spaces and over-sourced in fast food restaurants and chronic stressors (8,9). Although the described articles above went through the food environment in relation to zoning policies and obesity, there were quite a few overarching themes and deficits in current New York city policy that are described subsequently.

One significant aspect for obesity development is the food environment in which a child lives, grows up in and learns in. The food available near public schools are quite prevalent in New York City given the high density of fast-food restaurants and bodegas. These fast-food restaurants, as well as convenience stores, offer cheap and caloric dense foods that are easily available to school aged children and adolescents (10). As per Neckerman et al., (8), students who attend schools near fast food restaurants had a higher likelihood of obesity. However, surprisingly, the poverty status of students did not matter as much as ethnicity did. As shown in Elbel et al., (11), *non*-low-income Hispanic, Black and Asian students still lived near unhealthy corner stores, fast food restaurants and supermarkets compared to Caucasian students in New York City. The average NYC public school had approximately ten bodegas or corner stores within 400 m or 5-minute walk from school. This was seen more so in low income and Hispanic minority students, taking into consideration the population density around the public school, public transportation and commercial zoning laws (8). The likelihood of obesity was nearly 1.2% greater for students who went to schools near a bodega or corner café/store (10). Another contributing factor is unhealthy options in existing bodegas and restaurants near schools. As Neckerman et al., (8) hypothesized, it is likely effective to enhance the nutritional status of food options in existing food outlets if density of fast-food restaurants cannot be decreased (10). To be feasible, though, the prices of these healthier options need to be competitive with the unhealthy snack/food options present in the store (5). Unfortunately, the pricing differences are a separate problem in terms of access and purchase (12). This segues into the concept that New York City does not necessarily have food deserts but food swamps where energy dense foods/unhealthy snacks overwhelm healthier choices (5). Yes, there are multiple stores, bodegas and supermarkets for children and their respective families to choose from, but the actual choices these families have financial wise paints a different story.

We need to use zoning policy as a way to converge law and public health. A comprehensive understanding and examination of a child's food environment is key to determining what policies need to be implemented. By limiting the density of fast-food restaurants around urbanized schools, as well as in lower income neighborhoods, we can combat one of the many arms contributing to childhood obesity (3, 13). Schools near commercial land and urbanized areas are typically housed by low income and minority students and are engrained in separatist zoning policies (7). Ver Ploeg (5) stated, "Zoning rules, such as the amount of parking required for new businesses, could make it more costly to develop a new store. Small grocery stores or convenience stores may face lower rent and parking costs, but they may have a harder time accommodating equipment or space needed for fresh produce or perishable products..." (5). By acknowledging these policies, we can identify ways to modify and better close the gaps in food environment between ethnic and low socioeconomic groups.

It is important to note that under New York City's Health Code, "the Board of Health may enact, alter, amend or repeal any part of the Sanitary Code" (14). In this way, the Health Commissioner and Mayor addressed the obesity problem by issuing rules to chain restaurants for mandatory calorie counts on all foods and healthier food options in childcare centers (7). These rules were made in the hopes that children and their families would make smarter food choices (14). In both 2006 and 2009, there were two proposals made to restrict the locations of fast-food restaurants near New York City schools but, unfortunately, the proposals did not pass (10). All was not lost, though. The National Policy and Legal Analysis Network to Prevent Childhood Obesity (NPLAN) created a food zone ordinance by prohibiting *new* fast-food restaurants from being built near schools, playgrounds and recreational centers (15). This is tricky as we have previously discussed the dense prevalence of fast-food restaurants fueling obesity in inner city children. However, according to the New York Academy of Medicine, there are areas in Staten Island, Bronx and Queens that have less dense fast-food restaurants/bodegas.

This is interesting as Bronx and Queens, for the most part, have a lower income index compared to other parts of the five boroughs. This ordinance would be perfect for these lower income neighborhoods to protect the built food environment for vulnerable children.

Additionally, the city implemented a program called "Healthy bodegas and healthy bucks" to try and address the rising obesity epidemic, along with the Green Carts Program and Garden to Café (5, 12). By supplementing healthier, perishable items in corner stores and promoting/teaching how to sustain community gardens in school are some solutions to an otherwise complex problem. Another promising program is called the Food Retail Expansion to Support Health (FRESH) Program that directly incentivizes building fresh markets/supermarkets in otherwise underserved/disadvantaged neighborhoods via subsidized fresh produce (12, 15). Widespread education and lessons on healthy eating habits as well as calorie counting could reap tremendous benefits in children and if these lessons are brought home to families. It is equally important that education on healthy eating is made available along with affordable and nutritious foods in poorer income neighborhoods (5).

The other significant aspect for obesity development is the green spaces and recreational spaces in which children and their families play. Proximity to parks and green spaces allow for increased physical activity and exercise that can help combat obesity (16). However, green space and green space usage increases as "neighborhood disamenities" decreases. Disamenities include neighborhood violence, commercial waste/noxious land use and pollution, among others (13, 16). Attention to these factors can allow for children to safely access parks or playgrounds without fear. As described above, poorer neighborhoods with ethnic minority groups historically have disinvestment in their community rather than investment of "health promoting resources" (13). These resources include parks, sidewalks, walking trails and healthcare clinics (9). Deprivation of these resources lead to chronic stress early in childhood that propagates into tangible health conditions later in life (17, 18).

Children and families who live closer to recreational areas are more physically active in i.e., basketball and racquetball (13). One overarching concept that kept recurring in these articles is the concept of community capacity building. This includes commitment from various levels of power in the local government and its community members to actively combat against unfair zoning policies and work towards a healthier neighborhood for all (14, 17). Stages of assessment for community deficits, planning and implementation of solutions and evaluation of progress are key steps.

Obesity and unhealthy lifestyle habits are seen as a choice but largely are a sum of one's built environment and one's food environment. Governments use exclusionary zoning to wall off minority populations and lower socioeconomic brackets in particular locations to protect investments and increase property values (3, 19). As we have seen, this leads to separate wealthy areas that include residential, educational, and business districts. Legally, the courts have sided with municipalities to create their own zoning policies and city planning to best serve the community (as was described with New York City's first zoning laws). To illustrate this, Schroeder et al., created a simple hypothesis to essentially investigate the physiologic effects of exclusionary zoning (9). This was achieved by tracking changes in BMI z-scores in children/adolescents who moved from lower to higher poverty communities and vice versa. Those who transitioned to a higher poverty neighborhood were predicted to have higher obesity risk, and the risks were greater for female children/adolescents and male children. Despite the various health initiatives above, there is still this inherent drive to benefit the advantaged and disregard the disadvantaged (4, 20). These initiatives come after the fact of discriminatory planning and unequal development within New York City's neighborhoods. Thorton et al., (4), emphasized that periodic modification of zoning codes is incredibly significant to ensure economic, social and healthcare equality. By doing so, community members can actively engage in conversation with urban planners and lawmakers to enhance their quality of life and to increase access to clinics, food markets and recreational spaces (4).

Limitations

There are several limitations to this narrative review. Firstly, one of the major limitations was the narrow search criteria looking into zoning policies affecting childhood obesity in New York City. Using this narrow frame to investigate overarching zoning laws led to a relatively small number of studies to review and learn from. Although the concept of racial and ethnic separation via zoning policies is widespread nationally, there are nuances in this particular location that would make comparisons and generalizability somewhat difficult. New York City is a large, unique urban municipality that does not have certain elements like "food deserts" as other locations in the United States. As well, there is a higher density of fast-food restaurants in New York City compared to other urban settings in the country.

Additionally, the types of studies themselves have various limitations including lack of information on types of foods purchased by students in corner stores or bodegas near schools, as well as the patterns of food purchases in non-public schools to compare (nearly 20% of the total student population in New York City). Another significant limitation is the lack of studies included that look not only into corner stores and supermarkets by schools but also the sidewalk mobile food carts that are in close proximity to schools. This would also have a large impact on childhood obesity despite advances in healthy eating as these carts are cheap and offer energy dense foods.

Being a narrative review is inherently a limitation due to the qualitative nature of the data presented. For future studies, it would be beneficial to incorporate more statistical methods and Geographic Information Systems Mapping (GIS) into the review to have a better understanding of what factors play into childhood obesity besides income, green space, and the built environment.

Conclusion

The built environment is a tangible construction and consequence of zoning policies in New York City. However, the downstream effects of these policies lead to inherent chronic stress, increased obesity and poorer health outcomes for ethnic and financially poorer city residents and children. Social inequity lies hand in hand with the surrounding one grows up in. Children are the product of their environment and the product of the resources that are invested in them. By improving food sources, removing high energy snacks/drinks options near schools, increasing green space, and removing the barriers created by exclusionary zoning, we begin to incentivize a better future for children. Allowing community members and public health officials to work together with local governments to gradually change zoning policies would benefit all New York City residents. Altering the built environment towards equity leads to favorable alterations in the social determinants of health. Since New York City government controls public transportation, zoning rules, public school and hospital system policy, a stronger push is needed to transition from "equality in access to equity in access," (11). Those communities that are over-resourced on unhealthy foods, toxic land use, crime infestation and sparce green space should have the opportunity to create the ideal space to live in and for children to thrive in.

Childhood obesity is an incredibly complex interplay between various environmental and genetic factors. The reconstruction of new zoning policies to include inherent distances between schools and fast-food restaurants/bodegas, mandatory fresh fruit/vegetables with cheaper, competitive prices will hopefully turn the childhood obesity epidemic around. The final recommendation rests in the hands of educators, parents, and caregivers. The changes to healthy food and green space access, if made, are venues to a better lifestyle for children and their families. However, because minority and low-income neighborhood are accustomed to energy dense foods and sedentary lifestyles, there must be a shift in mentality to take advantage of newer, healthier resources. These approaches taken together will help reduce childhood obesity in New York City.

References

(1) Corburn J. Urban planning and health disparities: Implications for research and practice. Plann Pract Res 2005;20(2):111-26.

(2) Maantay J. Zoning, equity, and public health. Am J Public Health 2001;91(7):1033-41.

(3) Wilson S, Hutson M, Mujahid M. How planning and zoning contribute to inequitable development, neighborhood health and environmental injustice. Environ Justice 2008;1(4):211-6.

(4) Thornton RLJ, Greiner A, Fichtenberg CM, Feingold BJ, Ellen JM, Jennings JM. Achieving a healthy zoning policy in Baltimore: Results of a health impact assessment of the TransForm Baltimore zoning code rewrite. Public Health Rep 2013;128(Suppl 3):87-103.

(5) Ver Ploeg M. Access to affordable, nutritious food is limited in "food deserts." Econ Res Service USDA 2010;8(1):20-7.

(6) Centers for Disease Control and Prevention. Prevalence of childhood obesity in the United States. Atlanta, GA: CDC, 2022. URL: https://www.cdc.gov/obesity/data/childhood.html.

(7) Chen SE, Florax RJ. Zoning for health: The obesity epidemic and opportunities for local policy intervention. J Nutr 2010;140(6):1181-4.

(8) Neckerman KM, Bader MD, Richards CA, Purciel M, Quinn JW, Thomas JS, et al. Disparities in the food environments of New York City public schools. Am J Prev Med 2010;39(3):195-202.

(9) Schroeder K, Day S, Konty K, Dumenci L, Lipman T. The impact of change in neighborhood poverty on BMI trajectory of 37,544 New York City youth: A longitudinal study. BMC Public Health 2020;20(1):1676.

(10) Rummo PE, Wu E, McDermott ZT, Schwartz AE, Elbel B. Relationship between retail food outlets near public schools and adolescent obesity in New York City. Health Place 2020;65:102408.

(11) Elbel B, Tamura K, McDermott ZT, Duncan DT, Athens JK, Wu E, et al. Disparities in food access around homes and schools for New York City children. PLoS One 2019;14(6):e0217341.
(12) Limone O, Sanchez N. Mapping food deserts (and swamps) in Manhattan and the Bronx. Medium 2019. URL: https://medium.com/@olivialimone/mapping-food-deserts-and-swamps-in-manhattan-and-the-bronx-46c6d8fc0804.
(13) Sallis JF, Glanz K. Physical activity and food environments: Solutions to the obesity epidemic. Milbank Q 2009;87(1):123-54.
(14) Loureiro MI, Freudenberg N. Engaging municipalities in community capacity building for childhood obesity control in urban settings. Fam Pract 2012;29(1):24-30.
(15) New York Academy of Medicine. Zoning against unhealthy food sources in New York City. NYAM 2010;1-15.
(16) Weiss CC, Purciel M, Bader M, Quinn JW, Lovasi G, Neckerman KM, et al. Reconsidering access: Park facilities and neighborhood disamenities in New York City. J Urban Health 2011;88(2):297-310.
(17) Gordon-Larsen P, Nelson MC, Page P, Popkin BM. Inequality in the built environment underlies key health disparities in physical activity and obesity. Pediatrics 2006;117(2):417-24.
(18) Rundle A, Diez Roux AV, Free LM, Miller D, Neckerman KM, Weiss CC. The urban built environment and obesity in New York City: A multilevel analysis. Am J Health Promot 2007;21(4):326-34.
(19) Lopez RP, Hynes HP. Obesity, physical activity and the urban environment: Public health research needs. Environ Health 2006;5(25):1-10.
(20) Kazis N. Ending exclusionary zoning in New York City's suburbs. New York: NYU Furman Center, 2020:1-53. URL: https://furmancenter.org/files/Ending_Exclusionary_Zoning_in_New_York_Citys_Suburbs.pdf.

Chapter 7

A SWOT analysis of Chikungunya vaccination protocols

Meaghan Threadgill, MPH
Sara Elzibak, BSc
and Satesh Bidaisee*, DVM, MSPH, EdD
Public Health and Preventive Medicine, St George's University, Grenada, West Indies

Abstract

Chikungunya is a neglected tropical disease that is vector borne via mosquitos and which has high morbidity and increasing mortality. Long term effects of Chikungunya virus (CHIKV) include myocarditis, uveitis, retinitis, hepatitis, and neurological disorders. Therefore, vaccine development is imperative for the prevention of CHIKV-induced chronic illnesses. Interest in a vaccine for CHIKV has faltered due to a higher prevalence in developing countries rather than wealthier, developed nations. Other viruses such as Zika and yellow fever have approved vaccines. A literature review was conducted to determine the status of the development of potential vaccines against CHIKV. There is currently no available vaccines nor antivirals specific for CHIKV. However, many clinical trials of vaccine candidates have begun. SMART (specific, measurable, achievable, realistic, time-bound) objectives were included in the analysis to provide a baseline. PubMed and Google Scholar were utilized. Search terms included Chikungunya virus, vaccination, and SWOT (strengths, weaknesses, opportunities, threats) analysis. Current research has focused on various vaccination protocols but none are

* ***Correspondence:*** Satesh Bidaisee, Professor, Public Health and Preventive Medicine, St George's University, Grenada, West Indies. Email: sbidaisee@sgu.edu

In: Public Health: Intersection of Health, Humans, Animals and the Environment
Editors: Satesh Bidaisee and Joav Merrick
ISBN: 979-8-89530-443-3

currently available. Several modified vaccinia virus Ankara (modified smallpox) and live-attenuated vaccines have entered clinical trials. VLA1553 is in phase 3 clinical trials and VRC-CHKVLP059-00-VP has completed phase 2. MVA-E3E26KE1 has been proven to be highly efficacious in mouse models and will soon enter clinical trials. Chikungunya virus has been an epidemic issue for decades. Vaccination progress has been slow but is advancing. Opportunities exist in the realm of vaccine formulations to target vector-borne diseases. Strengths include a wealth of prior studies to implement a safe and effective vaccine protocol for CHIKV.

Introduction

Chikungunya virus (CHIKV) is a *togaviridae* alphavirus that has an enveloped positive-strand RNA virus coding for four nonstructural proteins and five structural proteins. The surface glycoproteins of CHIKV are potential targets for vaccines. Antibodies recognize these proteins during immunologic reactions to natural infection (1). As such, many vaccine candidates have been designed to mimic the immune response by targeting structural proteins. There are multiple genotypes/strains of CHIKV. Single mutations can alter the virus and may increase transmissibility (2). *Aedes (Ae) aegypti* and *Ae albopictus* are the primary vectors of CHIKV. Until recently it was difficult to distinguish between CHIKV and other viruses. The E1 and E2 surface proteins of CHIKV are involved in the infection process. Non-structural proteins may assist with viral replication. The incubation period of the virus is approximately 1-12 days and symptoms often occur three days post infection (3).

Although CHIKV is globally dispersed, much greater prevalence occurs in tropical areas (4). CHIKV is considered a neglected tropical disease and can be considered zoonotic since it also presents in non-human primates. *Ae aegypti* and *Ae albopictus* mosquitos are primarily responsible for the transmission of CHIKV (3). Arthritis and other chronic conditions such fever, rash, and neurological disorders can result from untreated CHIKV infections (5). Consequently, CHIKV has a high morbidity leading to high disability adjusted life years (DALYs) (6). In most cases disease duration of CHIKV is greater than other viruses, which further increases morbidity. Despite this virus having high morbidity, other mosquito-borne viral diseases such as Dengue have overshadowed the damage caused by CHIKV (5). Wildlife contributes to the dissemination of CHIKV. Non-human primates can serve as a host for CHIKV allowing it to survive in many populations (7).

CHIKV was first discovered in the 1950s in Tanzania. The first major outbreaks were during the 1960s and 70s in Africa and Asia. Since then, the virus has spread across the globe and is present in almost every continent (1). In 2004 there was a large outbreak in Kenya followed by outbreaks in Comoros and La Reunión in 2005. From 2013-2016 outbreaks occurred in the Caribbean and Americas. CHIKV has subsequently infiltrated more than 100 countries (3).

The purpose of this paper was to determine the strengths, weaknesses, opportunities, and threats when trying to formulate and distribute a vaccine for Chikungunya virus (CHIKV).

Our review

A literature review of current research was conducted, followed by an analysis of the status of vaccine protocols.

Research studies were selected using Google Scholar and PubMed databases. Boolean operators were used including "chikungunya" AND "vaccine" in addition to phrases such as "Chikungunya vaccine," "vaccine for chikungunya." From PubMed and Google Scholar 972 articles were retrieved.

Of the 972 articles retrieved, 28 were selected for inclusion since they were relevant, research-based, and within the last ten years.

A SWOT (strengths, weaknesses, opportunities, threats) analysis was performed using the scientific papers identified in the search. The SMART criteria (specific, measurable, achievable, realistic, time-bound) were utilized as the backbone for the analysis. Specificity focused on Chikungunya vaccine candidates. Vaccine formulations were measured based on efficacy, safety, and clinical trial status.

Comparison charts were included. Cost-benefit analysis would vary deepening on each Nation's economic status and CHIKV prevalence.

Findings

Accurate Chikungunya testing and diagnosis can be difficult because of symptom similarity to other vector borne diseases (8, 9). Consequently, molecular tests that are rapid and can accurately distinguish between Chikungunya, Dengue, and Zika viruses should be developed and implemented. ELISA and PCR tests are feasible for distinguishing between viruses. Reverse transcription polymerase chain reaction (RT-PCR) has been used to test populations and differentiate between different viruses (10), but this method is costly and time consuming. There are currently no approved CHIKV vaccines, and antivirals for CHIKV are limited (11). Vaccination for this vector-borne disease could greatly reduce its negative impact on health and improve the quality of life of those living in areas at risk of infection.

Since its discovery, Chikungunya has caused several epidemics across the globe. CHIKV was first detected in the early 1950's in Tanzania and has since spread worldwide (11). Chikungunya cycles between mosquitos, water sources, humans, and non-human primates. Monkeys are believed to be an important reservoir for CHIKV (7). Therefore, it is important to mitigate each level of the transmission cycle such as a) reducing standing water for mosquito breeding and b) monitoring wildlife for CHIKV prevalence.

Many symptoms of CHIKV can be recurrent causing chronic disability. There is minimal information related to the economic costs of the virus but there is documentation that the treatment and job loss has tremendous impact on communities. India has the highest caseload likely due to climate and dense population. There has been an increase in annual cases (12). India also has an unemployment rate of about 8%, which is lower than the global average (13). Consequently, research on vaccinations and anti-virals should be emphasized.

Research has been continuous but localized to specific outbreaks. Studies have included symptoms, prevention, potential vaccination protocols, and economic impacts (7, 14). Public health is centered around primary prevention, and thus it is essential to complement these efforts with vaccines for infectious diseases. Vector control is not a fool-proof method, thus other preventative and direct treatment methods will still be needed. Research is being conducted for a variety of vaccination methods. Virus-vector vaccines have appeared to be the most promising thus far, although none are commercially available (15). Factors that need to be evaluated before wide-spread use include safety, efficacy, and financial resources for distribution to low-income areas. CHIKV has been known to mutate rapidly and is highly virulent, which can complicate vaccine development. Each method of vaccination needs to be analyzed and compared (6).

The Pan American Health Organization performed a SWOT analysis on integrated management strategies for disease control (16). Results of this analysis highlighted lack of resources and training for personnel as a weakness to controlling vectors. Potential threats were climate change, globalization, societal inequalities, and community perception. Strengths included political support and consistent surveillance. Opportunities were demonstrated with the advancement of technology to better trace disease spread (16). A current threat is the ability of the virus to adapt to vectors and hosts. It was initially thought that the virus was primarily transmitted by *Ae.aegypti* mosquitos however the virus has been able to survive in other species such as *Ae. albopictus*. This could contribute to a greater dispersion of the virus and increased damage. There is weakness in the under-reporting of cases. Current vaccine platforms include formalin-inactivated, live-attenuated, viral-vectored, and nucleic acid-base vaccines. The live-attenuated vaccines tend to be derived from molecular clones and have been effective in mouse studies. Viral-vectored vaccines have been tested in non-human primates with promising results.

One platform combines Eilat, an insect specific virus, with CHIKV. A single dose of this formula is able to initiate an immune response. During phase 1 of clinical trials the viral-vectored vaccine was given in three doses, with this formulation recently beginning phase 2 trials. Nucleic acid-base vaccines tend to be easier to produce but are less effective. There have been numerous obstacles delaying the production of a Chikungunya vaccine (17).

Table 1. SWOT analysis of current vaccine status

Strengths	Weaknesses	Opportunities	Threats
• Many years of study • Vaccine platforms already being studied • Clinical trials are progressing	• Underestimate of case-fatality • Lack of funding	• Continue Clinical trials of current vaccine formulations • Potential for provisional license of vaccine • Ability to form international partnerships to prevent neglected tropical diseases	• Life-long disability and complications due to CHIKV. • Risk to lower income nations • CHIKV mutations

Research showed that CHIKV has been a global issue since 2004, although the virus itself was first discovered in the 1950's. This situation describes a need for a vaccination protocol. A vaccine must be specific for this virus and its various mutations. CHIKV spreads rapidly when outbreaks occur, so prevention is the best method (6). There are already some vaccine formulas being tested. Proposed formulas have been proven to be safe and moderately effective in non-human primates. A measles-vectored vaccine was given to macaques and then they were challenged with a contagious dose of CHIKV. The vaccinated individuals did not demonstrate clinical symptoms of the virus (15).

The strengths of a recombinant measles Chikungunya vaccine includes a potent immune response with minimal side-effects (4). Weaknesses of the vaccination programs directed toward CHIKV are the lack of progress made, despite many years since the discovery of the virus. Opportunities include the fact that there are currently no commercially available vaccines for CHIKV. Threats are the mutation rate of the virus and lack of funding for the research of the disease (8). Additionally, CHICKV is zoonotic and can present in non-human primates (7). In conclusion, a priority must be placed on vaccine development and testing, as well as production.

Table 2. Recombinant MVA (modified vaccinia virus Ankara) vaccine candidates against CHIKV

Name	Doses	Effects	Limitations
MVA-CHIKV (C/E3/E2/6K/E1)	1-2	Production of IgG antibodies, protection against CHIKV in mice and non-human primates	N/A
MVA-CHIK	2	Produces anti-CHIKV antibodies in mice	Low levels of neutralizing antibodies
MVA-E3E2 (E3/E2)	1	Protects CHIKV challenged mice	Low levels of neutralizing antibodies
MVA-6KE1 (6K/E1)	1	Partial protection of lethal CHIKV dose in mice	Infectious CHIKV found in some organs
MVA-E3E26KE1	1	High antibody production, protection from infection in mice	N/A
MVA-CHIKV- sAB^+	4	Reduced viral titer	CHIKV titer present until day 4

Source: Adapted from García-Arriaza J, Esteban M, López D. Modified vaccinia virus ankara as a viral vector for vaccine candidates against chikungunya virus (21).

Non-live and inactivated vaccines usually require more doses due to lower potency. Live-attenuated vaccines are commonly used for many viruses (18). Huckle et al., 2021 identified that while macrophages tend to be reservoirs of the virus during late-stage infections, dermal fibroblasts are the initial target following an infected bite (19). CHIKV and many Chikungunya vaccine platforms illicit a strong IgG3 response (20). Live-attenuated vaccines may cause higher rates of symptoms. VLA1553 is currently starting phase 3 clinical trials but previous live-attenuated vaccines were discontinued due to safety concerns. An inactivated vaccine BBV87 complete phase 1 trials using a 2-dose regime. The US National Institutes of Health helped develop a virus like particle platform which completed phase 2 of clinical trials. Many recombinant vaccines are also showing promising progress (19).

Table 3. CHIKV vaccine candidates in development

Name	Affiliation	Method	Status
TSI-GSD-218 (181/clone25)	US Army Medical Research Institute of Infectious Diseases, University of Maryland	Live-attenuated CHIKV strain	Phase 2 Discontinued due to safety uncertainty
VLA1553	Valneva, Austria	Live-attenuated CHIKV with nsP3 deletion	Phase 3
BBV87	Bharat Biotech (BBIL)	inactivated whole virion vaccine	Completed phase 1
VRC-CHKVLP059-00-VP	US National Institutes of Health, PaxVax	Virus-Like Particle Vaccine (VLP)	Completed phase 2 clinical trials
CHIKV/IRES	University of Texas Medical Branch, Takeda Pharmaceuticals	Recombinant modified CHIKV	Might begin clinical trials
MV-CHIK	Institute Pasteur, Themis Bioscience	Recombinant live-attenuated measles vaccine with CHIKV proteins	In Phase 2
ChAdOx1 Chik	University of Oxford	adenoviral vector unable to replicate	In phase 1
VAL-181388	Moderna Therapeutics Inc	mRNA	Recruiting for phase 1

Source: Adapted from Hucke FIL, Bestehorn-Willmann M, Bugert JJ. Prophylactic strategies to control chikungunya virus infection (19).

A modified vaccinia virus Ankara was developed to elicit innate and humoral immune response against CHIKV. Multiple vaccines have been tested with the MVA vector but most of them require multiple doses (21). Chikungunya virus can also be transmitted vertically from mothers to infants so perhaps vaccination could mitigate this (19).

Limitations

Achievability of a commercial vaccine is dependent on continuing research as well as funding for trials. To realistically formulate a CHIKV vaccine, more trials may need to be conducted. Many CHIKV vaccines have not made it past phase 2 of human trials due to inconsistent interest (17). Lack of funding and difficulty determining efficacy has stalled the progress of some vaccines. The efficacy of the vaccine relies on the formulation method and dosage. Vaccine platforms might be expedited in the future due to the COVID-19 pandemic since the pandemic provided more efficient avenues for vaccine approval and proved the use of newer RNA-mediated vaccines (22).

Conclusion

There are numerous benefits to finalizing a licensed vaccine against Chikungunya virus. The best method will depend on efficacy, safety, and resources. The cost should be made reasonable for endemic areas. Each platform has their own unique benefits and challenges, but they show promising results.

Recommendations

The implementation of a CHIKV vaccine would also reduce the need of flogging and prevent exposure to harmful chemicals. Chikungunya virus has high morbidity and greatly impacts communities, which could easily be prevented through vaccination. Resources should be provided to expedite trials to produce a bulk available vaccine. Emphasis should be placed on the development of vaccinations to prevent disease.

References

(1) Goyal M, Chauhan A, Goyal V, Jaiswal N, Singh S, Singh M. Recent development in the strategies projected for chikungunya vaccine in humans. Drug Des Devel Ther 2018;12:4195-206.

(2) Higgs S, Vanlandingham D. Chikungunya virus and its mosquito vectors. Vector Borne Zoonotic Dis 2015;15(4):231-40.

(3) Vairo F, Haider N, Kock R, Ntoumi F, Ippolito G, Zumla A. Chikungunya: Epidemiology, pathogenesis, clinical features, management, and prevention. Infect Dis Clin North Am 2019;33(4):1003-25.

(4) Ramsauer K, Schwameis M, Firbas C, Müllner M, Putnak RJ, Thomas SJ, et al. Immunogenicity, safety, and tolerability of a recombinant measles-virus-based chikungunya vaccine: A randomized, double-blind, placebo-controlled, active-comparator, first-in-man trial. Lancet Infect Dis 2015;15(5):519-27.

(5) Alvarado LI, Lorenzi OD, Torres-Velásquez BC, Sharp TM, Vargas L, Muñoz-Jordán JL, et al. Distinguishing patients with laboratory-confirmed chikungunya from dengue and other acute febrile illnesses, Puerto Rico, 2012-2015. PLoS Negl Trop Dis 2019;13(7):e0007562.

(6) Powers AM. Vaccine and therapeutic options to control chikungunya virus. Clin Microbiol Rev 2017;31(1):e00104-16.

(7) Althouse BM, Guerbois M, Cummings DAT, Diop OM, Faye O, Faye A, et al. Role of monkeys in the sylvatic cycle of chikungunya virus in Senegal. Nat Commun 2018;9(1):1046.

(8) Sam IC, Kümmerer BM, Chan YF, Roques P, Drosten C, AbuBakar S. Updates on chikungunya epidemiology, clinical disease, and diagnostics. Vector Borne Zoonotic Dis 2015;15(4):223-30.

(9) World Health Organization. Chikungunya. WHO 2020. URL: https://www.who.int/en/news-room/fact-sheets/detail/chikungunya.

(10) Valentine MJ, Murdock CC, Kelly PJ. Sylvatic cycles of arboviruses in non-human primates. Parasit Vectors 2019;12(1):463.

(11) Ahola T, Couderc T, Ng LF, Hallengärd D, Powers A, Lecuit M, et al. Therapeutics and vaccines against chikungunya virus. Vector Borne Zoonotic Dis 2015;15(4):250-7.

(12) European Centre for Disease Prevention and Control. Chikungunya worldwide overview. ECDC 2021. URL: https://www.ecdc.europa.eu/en/chikungunya-monthly.

(13) Center for Monitoring Indian Economy. Unemployment Rate in India. CMIE 2021. URL: https://unemploymentinindia.cmie.com/.

(14) Garcia A, Diego L, Judith B. New approaches to chikungunya virus vaccine development. Recent Pat Inflamm Allergy Drug Discov 2015;9(1):31-7.

(15) Rossi SL, Comer JE, Wang E, Azar SR, Lawrence WS, Plante JA, et al. Immunogenicity and efficacy of a measles virus-vectored chikungunya vaccine in nonhuman primates. J Infect Dis 2019;220(5):735-42.

(16) World Health Organization. Integrated management strategy for arboviral disease prevention and control in the Americas. PAHO 2020. URL: https://iris.paho.org/bitstream/handle/10665.2/52492/9789275120491_eng.pdf?sequence=1&isAllowed=y.

(17) Rezza G, Weaver SC. Chikungunya as a paradigm for emerging viral diseases: Evaluating disease impact and hurdles to vaccine development. PLoS Negl Trop Dis 2019;13(1):e0006919.

(18) Vetter V, Denizer G, Friedland LR, Krishnan J, Shapiro M. Understanding modern-day vaccines: What you need to know. Ann Med 2018;50(2):110-20.

(19) Hucke FIL, Bestehorn-Willmann M, Bugert JJ. Prophylactic strategies to control chikungunya virus infection. Virus Genes 2021;57(2):133-50.

(20) Tschismarov R, Zellweger RM, Koh MJ, Leong YS, Low JG, Ooi EE, et al. Antibody effector analysis of prime versus prime-boost immunizations with a recombinant measles-vectored chikungunya virus vaccine. JCI Insight 2021;6(21):e151095.

(21) García-Arriaza J, Esteban M, López D. Modified vaccinia virus ankara as a viral vector for vaccine candidates against chikungunya virus. Biomedicines 2021;9(9):1122.

(22) Trovato M, Sartorius R, D'Apice L, Manco R, De Berardinis P. Viral emerging diseases: Challenges in developing vaccination strategies. Front Immunol 2020;11:2130.

(23) American Veterinary Medical Association. (2019). Disease precautions CHIKV. AVMA 2019. URL: https://www.avma.org/resources/public-health/disease-precautions-hunters#chikungunya.

(24) Gao S, Song S, Zhang L. Recent progress in vaccine development against chikungunya virus. Front Microbiol 2019;10:2881.

(25) Gerke C, Frantz PN, Ramsauer K, Tangy F. Measles-vectored vaccine approaches against viral infections: A focus on chikungunya. Expert Rev Vaccines 2019;18(4):393-403.

(26) Giel-Moloney M, Goncalvez AP, Catalan J, Lecouturier V, Girerd-Chambaz Y, Diaz F, et al. Chimeric yellow fever 17D-Zika virus (ChimeriVax-Zika) as a live-attenuated Zika virus vaccine. Sci Rep 2018;8(1):13206.

(27) Juarez JG, Garcia-Luna S, Chaves LF, Carbajal E, Valdez E, Avila C, et al. Dispersal of female and male Aedes aegypti from discarded container habitats using a stable isotope mark-capture study design in South Texas. Sci Rep 2020;10(1):6803.

(28) Naranjo DP, Qualls WA, Jurado H, Perez JC, Xue RD, Gomez E, et al. Vector control programs in Saint Johns County, Florida and Guayas, Ecuador: Successes and barriers to integrated vector management. BMC Public Health 2014;14(1):674.

Chapter 8

The impact of antimicrobial resistance: Antimicrobial stewardship

Samuel Ruch, DVM, MPH
Sara Elzibak, BSc
and Satesh Bidaisee*, DVM, MSPH, EdD
Public Health and Preventive Medicine, St George's University, Grenada, West Indies

Abstract

Antimicrobial resistance (AMR) poses a serious threat to human and animal life on earth. Policies that are implemented to slow down or reverse AMR are termed antimicrobial stewardship programs (ASPs). The focus of this narrative review is to determine how ASPs policies can effectively change the way that antimicrobials are used. Specific search terms were used across three databases to generate an initial pool of papers. After careful review of all search results, 27 were selected and analyzed according to the type of study, themes, and discussion of AMR and ASPs. There was variety in the types of studies analyzed, however, there was a striking lack of research in veterinary medicine. Major themes of ASP success were discovered as follows: interprofessional collaboration, dedicated stakeholders, and importance of education. Several barriers to ASP success are discussed from staffing to facilities to information technology. Low to middle income countries (LMICs) have additional barriers to ASP implementation. ASPs can be effective methods at reducing antimicrobial usage and cost of treatment while

---

* ***Correspondence:*** Satesh Bidaisee, Professor, Public Health and Preventive Medicine, St George's University, Grenada, West Indies. Email: sbidaisee@sgu.edu

In: Public Health: Intersection of Health, Humans, Animals and the Environment
Editors: Satesh Bidaisee and Joav Merrick
ISBN: 979-8-89530-443-3

improving patient care; however, current research on ASPs has shown little impact on AMR.

Keywords: antimicrobial resistance, AMR, antibiotic resistance, AR, antimicrobial stewardship programs, ASPs, public health, interprofessional collaboration, one health

Introduction

Antimicrobial resistance (AMR) poses a major public health threat to human and animal life on earth. The World Health Organization (WHO) in their World Health Statistics 2022 report lists that AMR is in the top 10 of all global threats to public health (1). The Centers for Disease Control and Prevention (CDC), in their 2019 Antibiotic Resistance (AR) Threats Report, revealed that the number of deaths due to resistant infections has decreased by about 18% since 2013; however, over 2.8 million people are infected and more than 35,000 people die each year from them in the United States (see appendix A) (2). As microbes become increasingly resistant to common treatments, infections will become more difficult to treat, resulting in increases in morbidity and mortality in humans and animals.

Antimicrobials are used on a daily basis in both human and animal species, and AMR is a serious concern worldwide. This is cause for alarm as new antimicrobials are desperately needed to address this issue; however, unless current antimicrobial practices are changed, future treatment options will face the same fate. Without policies that hold all stakeholders accountable, AMR will continue to be a global issue.

Antimicrobial stewardship programs (ASPs) are important programs in the fight against AMR with the main goals of providing quality patient care, only using antimicrobials when necessary, working to decrease levels of AMR, and lowering costs associated with care (3).

The focus of this narrative review is to determine how ASPs policies can function effectively to reduce the impact of AMR, discover key factors that resulted in ASP success, and ultimately assess the strengths and weaknesses of the current research on ASPs worldwide. The aim is to answer the following question: How can ASPs effectively change the way that antimicrobials are being used?

Methods

There has been consistent research on the topics of AMR and ASPs for several years which has resulted in governmental and organizational movements to prevent the spread of resistance. It is no longer a question of whether ASPs should be implemented, but rather, what types of stewardship programs are most effective and produce the greatest impact on AMR, patient outcomes, and cost of care.

Database searching for this narrative review took place from May 2020 to July 2022. Specific search terms were used and analyzed across three databases: PubMed, DOAJ and Academic Search Complete. The terms used for the search were "antimicrobial drug resistance," "antimicrobial stewardship," and "stewardship program," which yielded 348 possible results.

The 348 papers were imported into Zotero reference management software. The full text of 54 articles could not be acquired leaving only 294 that were initially screened. Following a careful review of titles and abstracts on Zotero, 87 papers met the initial screening criteria. Then a full assessment of these papers ensued, resulting in the final 27 that were included in the study (see Appendix B for Study Selection Flowchart).

To be included in this narrative review, sources needed to be from an academic journal with full free text available in English, published from 2018-2022, with a focus on the impacts and costs of AMR as well as planning, effectiveness, impact evaluation or attitudes and beliefs surrounding ASPs.

Sources were excluded if they were narrative review articles, newspaper articles, book reviews, dissertations, theses or if the studies focused on one specific disease or infection.

These papers were then analyzed by type of study, field of study, themes, purpose and discussion of AMR and ASPs. This was done to determine how ASPs are being researched, discover barriers that hinder implementing ASPs, and evaluate the attitudes of key stakeholders surrounding AMR and ASPs. The results were then interpreted to present their overall effectiveness as well as the strengths and limitations of the research.

Results

Of the 27 articles that were fully assessed in this review, 37% (10/27) were cross-sectional, 2 of which were prospective. These cross-sectional studies were mainly conducted through surveys and included analysis of knowledge, attitudes and perceptions of antimicrobials, AMR and ASP. Observational studies comprised 29.6% (8/27) of the articles with five of the studies being retrospective. Case studies made up 14.8% (3/27) of the articles, with one study including two separate case studies to be analyzed. Cohort studies were 7.4% (2/27) of the articles, and one mixed methods study accounted for 3.7% (1/27). Finally, one systematic review and meta-analysis was included and accounted for 3.7% (1/27) of the studies analyzed (see Figure 1). The systematic review was performed on 13 studies, with 10 studies showing successful ASP interventions (4). Of the analyzed articles, the majority discussed the role of AMR and ASPs in the hospital, ICU, tertiary care facilities, and primary care centers, while others assessed the knowledge and attitudes surrounding ASPs and AMR in clinical students and pharmacists. It is important to note that only one article was specifically focused on veterinary medicine and the impacts of ASP in the veterinary field.

Table 1. Successful ASP themes and takeaways

Themes	Major takeaways
Interprofessional collaboration	• Nurses, pharmacists and physicians working together reduced use of antibiotics, improved patient care and lowered patient cost (5). • Collaborations between professionals created the environment for interprofessional learning (5). • Clinicians and pharmacists reviewing antimicrobial therapy options together resulted in decreased antimicrobial use without impacting patient outcome (12). • Collaborations with local stakeholders (patients & public) to develop understandings of cultural and behavioral factors affecting AMR (13).
Dedicated stakeholders	• Financial support, hiring specialists, encouragement to participate in educational events were associated with better ASP results (6). • Deep awareness and collaboration of all national professional stakeholders led to success of ASP program in swine (7). • Increased leadership support and involvement leads to improved collaboration within the ASP (8).
Importance of proper education	• Knowledge of antimicrobials, their spectrum of activity and how it relates to AMR has been found to reduce AMR (15). • Proper training and education of new prescribers will improve positive health outcomes (14). • Educational opportunities allow for those in the ASP to gain knowledge and become future leaders in the program (19).

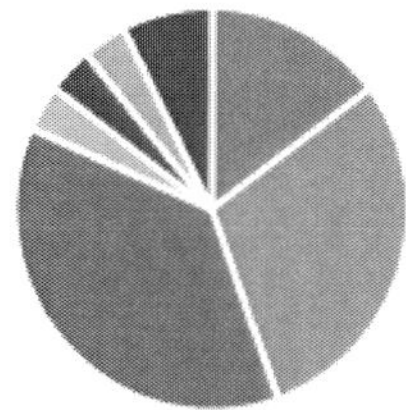

Figure 1. Types of studies analyzed.

Throughout the research process there were several factors identified that lead to ASP success in their studies. Interprofessional collaboration, dedicated stakeholders, proper education, and oversight by specialized professionals were all seen as positive indicators of ASP success and were consistently mentioned throughout the literature.

Discussion

A major theme that is frequently discussed throughout the selected articles is the importance of collaboration of all stakeholders within the ASP. A One Health approach with collaboration and communication between multiple sectors is required in the battle against AMR for better outcomes (5). In a study performed by Alghamdi et al., (6) on ASP implementation in a Saudi hospital, the decision to integrate an ASP, that was initiated from the top officials at the hospital, was supported financially by management and expanded the workforce to include infectious disease clinicians and pharmacists, all while managers at the hospital encouraged staff to actively engage in the educational events provided; all of which resulted in a decreased usage of antimicrobials (6). Verliat et al., (7) reported on the effectiveness of an ASP program in swine production and found that deep involvement and awareness of all national professional stakeholders was the basis of their ASP. This allowed for identifying and understanding the problem, implementing proper usage guidelines, and communicating among all stakeholders which ultimately enabled the program's success at reducing antimicrobial usage in swine while preserving the efficacy of the antimicrobials (7). Houng et al., (8) revealed that increased levels of leadership support and dedication to the ASP lead to more engagement in the activities by team members and improved collaboration between all stakeholders.

Support from top management and major stakeholders is the first step to take to ensure ASP success, but while the support from the top is important, it only works if managers, clinicians, pharmacists, nurses, and staff are actively collaborating with each other. In a study performed by Stephan Schmid et al. that focused on interprofessional collaboration, they showed that the active cooperation of nurses, pharmacists and physicians reduced the overall use of antibiotics and the use of broad-spectrum antibiotics, ultimately improving patient care and resulted in lowered costs (5). Other articles specifically discussed adding an infectious disease consultant or a clinical pharmacist to the ASP team to collaborate on treatment decisions, contrasting to other studies where the entire ASP team was involved in the decision on antimicrobial use (9-11). A novel approach to this collaboration is the use of a modified prospective audit and feedback (PAF) system that was demonstrated by Kim et al., (12). It involved both clinicians and pharmacists reviewing choices of antimicrobials rather than having a pre-determined selection of antimicrobials. This resulted in a decreased antimicrobial usage without negative effects on the patients (12). Veepanattu et al., (13). discussed the importance of building collaborations with local stakeholders; for example, the inclusion of the public and the patients as part of the research can provide a better understanding of cultural and behavioral factors that may negatively affect the impact of ASPs. These authors further discussed how organizations, funding bodies, and research institutions needed to create strategic partnerships to strengthen the research dedicated to AMR (13).

In contrast to the success of collaboration and communication of all stakeholders, a study performed by Sarwar et al., (14) found that the majority of pharmacists did not communicate with prescribers concerning the uncertainty of a prescription, and that the lack of communication was a major factor for poor antimicrobial usage according to their study (14).

Many of the studies reviewed in this paper highlighted the role that proper and continued education plays a role in the success of all stewardship programs. Knowledge of antimicrobials, their spectrum of activity, and how their spectrum relates to resistance have been found to aid prescribers when selecting the correct antimicrobial, and therefore, reduce AMR (15). Without quality education that details how microbes become resistant to medications, simple ways to prevent resistant organisms, and a wide array of clinical cases to practice the skill of antimicrobial selection, these programs cannot succeed. Nasr et al., (16) investigated the knowledge of students with respect to antimicrobial stewardship cases and the findings suggest that education on clinical reasoning and decision making should be implemented into educational curriculum for pharmacy students (16). Lack of knowledge in any of these areas for prescribers can lead to uncertainty about antimicrobial selection and ultimately improper antimicrobial usage. Providing proper education and training of new prescribers promoted positive health outcomes for patients (14). Overall, improved education on antimicrobials has worked to slow down the process of AMR around the world (15).

In a qualitative study performed by Sundvall et al., (17), many participants including doctors and nurses, highlighted the need for regular education and treatment guidelines to aid in diagnosis and treatment of infections. These authors discussed how practical experience and skills may not be enough to combat AMR; and wisdom and communication among colleagues may provide better recognition of the best treatment options (17). This belief is also shared by Schmid et al., (5) as they discussed the role of collaboration between professionals of their respective field as a major key to success of ASPs because it created the optimum scenario where experts in their respective fields are involved in complex case management which fosters an environment that allows for interprofessional learning. A randomized trial of ASP, done by McIsaac et al., (18), found that education, clinical decision aids, and audit and feedback mechanisms along with compensation for ASP activities resulted in decreased antibiotic prescriptions for infections such as respiratory and urinary diseases as well as a reduced antimicrobial treatment duration (18).

An added benefit of education for ASPs was discussed, in the study performed by Kishida and Nishiura (19), and showed that continued educational opportunities for the participants of ASPs allowed staff to gain better antimicrobial stewardship knowledge, but more importantly, acquire the tools to become future leaders in the program (19).

Conversely, a lack of education about AMR and antimicrobials leads to irrational prescription and is associated with the spread of AMR (20).

In general, there are a multitude of factors that hinder the success of ASP programs and ultimately negatively impact the rate of resistance around the world. Indiscriminate use of antimicrobials in animals and humans has created medical complications in patients, caused unnecessary expenses, as well as propagated the development of resistance (21). Some authors commented that diagnostic uncertainty, improper choice and incorrect duration of antimicrobial therapy, and lack of surveillance to monitor antimicrobial use are major components of inappropriate use of antimicrobials (22). Others discussed how the ease of antimicrobial availability to the public without a prescription severely and negatively impacts the success of ASPs (14).

Other barriers to success of ASP programs include: shortage of ASP staff, incompatible IT systems, lack of microbiology personnel and facilities, and physician resistance to ASP formulary restrictions and policies (6). Additionally, some authors commented on the lack of financial and human resources, the lack of knowledge on prescribing practices and the lack of cooperation from prescribers also were contributing factors as barriers to ASPs (23). Even more studies echoed these sentiments as it was also reported that a lack of ASP activities or activities performed in unstructured manner also prevented ASP success (24). However, Shin et al., (25) argued that improvements in ASP may not correspond to the number of ASP activities because variation in quality of activities are likely to produce different results and not have equal success.

Low and middle income countries (LMICs) have additional barriers to successful implementation of ASPs in the form of infrastructure, surveillance and lack of capacity. In general, these barriers are multifactorial and include but are not limited to: low vaccination rates, higher infectious disease caseload, unavailability of antimicrobials, poor laboratory utilization or availability, and lack of laboratory equipment (23, 24). One common theme shared by many other authors stresses the lack of access to microbiology services and diagnostic tools result in increased AMR burden of LMICs and increased barriers to success (20). In summary, limited access to diagnostics, absence of supporting infrastructure, unavailability of the correct drugs, and an overall lack of knowledge of the healthcare staff are identified as the major problems for LMICs (8).

Limitations

Current research on the impact and effectiveness of ASPs has been consistently variable. While many studies have been effective at demonstrating reduction of costs associated with care and reduction of total antimicrobial usage, the research is deficient at demonstrating the reduction of AMR after implementing the ASPs. Tartof et al., (26) discussed how single target studies can demonstrate a decrease in resistance to a single antimicrobial; however, this is usually compensated by the increased use of another antimicrobial, otherwise known as the "squeezing the balloon" phenomenon. Additionally, ASP research has yet to show consistent statistically significant reductions in AMR. According to Diaz-Madriz et al., (21), the effects from ASP interventions usually take approximately five years to manifest. As such, the lack of demonstrating successful reduction of AMR could potentially be due to a short study duration. It is also important to address that some studies may show positive responses to implementation of an ASP, but it is impossible to say that the improvement is a direct result of the ASP alone. There are several external factors that may play a role in positively influencing the study findings. For example, increased awareness of the impact of AMR in healthcare facilities and improved hygiene practices such as during COVID-19 pandemic could have improved ASP results.

A potential limitation of this narrative review was that this research was taking place as the SARS-CoV-2 (COVID-19) pandemic was impacting the globe. The COVID-19 pandemic resulted in significantly lost progress in the battle against AMR. According to the CDCs COVID-19 AR Special Report, reductions in AMR deaths were continuing until 2020, but the pandemic fueled more resistant infections and increased antimicrobial usage while simultaneously provided less data collection and preventative measures to combat the spread of resistance (27). The increased use of antimicrobials during the COVID-19 pandemic may have played a role in affecting some of the studies that were conducted. As a result, these studies may have reported inconclusive or poor results after implementation of ASPs. Tomcyzk et al., (28) concluded that there was an increase in antibiotic prescriptions during the pandemic and that this also was a common finding in LMICs.

Another limitation of this review was not acquiring access to the 54 articles; some of these articles may have filled in gaps or discussed alternatives not found in the papers in this study. Finally, the majority of the articles examined in this paper were focused on human medicine and only one paper was specifically focused on the veterinary industry. In the initial database search, none of the search terms included veterinary medicine or animals, which may have resulted in missed opportunities to find specific veterinary programs that may have been successful as well as provided keen insights on the success of these programs.

Recommendations

For ASPs to be effective there needs to be extreme dedication to the program from all stakeholders. A multidisciplinary approach that is structured on clinical leadership with motivated and enthusiastic participants are crucial elements to the success of ASPs (29). Commitment from governmental agencies, the public health sector, healthcare facilities, veterinarians, physicians, pharmacists, nurses and the general public is needed for these programs to be successful. The potential barriers and challenges that the program may encounter need to be addressed during the planning phase of ASPs by discussing ways to overcome the recognized challenges. In addition, educational opportunities, such as weekly rounds on antimicrobial usage and resistance, for all staff members are crucial to the success of ASPs long-term as it enables participants to become leaders at their facility. These opportunities need to be actively encouraged and/or required by stakeholders and utilized to the fullest extent by staff members.

Although many different forms of ASPs are used, in the Shirazi et al., (4) systematic review and meta-analysis of ASPs, it was discovered that the most commonly used methods of ASP were prospective audit and feedback (PAF) and formulary restrictive, followed by educative and guideline development. Ultimately, these programs provided evidence of effectively reducing the misuse of antimicrobials that lead to resistance as well as decreasing the cost of antimicrobial treatment that resulted in decreased rates of AMR (4). To further assist these types of ASP methodologies, Bolten et al., (30) discussed the addition of an automatic discontinuation of antibiotics policy (ADAP) and argued that the implementation of this change caused a shift from a passive ASP into a more active, effective and efficient model of ASP. In summary, the models that have proven to be the most effective as well as the best chance of successfully decreasing AMR include: PAF, formulary restrictive and ADAP.

The importance of the role of education in the success of ASPs is paramount. Ensuring all prescribers and pharmacists have knowledge on antimicrobial use is critical to curb inappropriate use of antimicrobials (31). For example, providing proper education about antimicrobials and mechanisms of resistance via clinical cases or seminars delivered by experts will give clinicians the confidence to accurately diagnose and treat patients.

Recommendations for future researchers are to follow and monitor ASPs for at least five years to truly acquire accurate data regarding the impact of ASPs, specifically on the reduction of antimicrobial usage and the final impact on AMR. Finally, all future research should utilize a One Health approach to the problem of AMR because it prioritizes gaining key insight from all major stakeholders.

Conclusion

In their current capacity, ASPs are not doing enough to combat AMR and prevent unnecessary use of antimicrobials around the globe. In the era of One Health policies being implemented worldwide, it is alarming that out of the twenty-seven papers that were analyzed in this report, only one study covered ASPs in veterinary medicine. Failing to coordinate with all major stakeholders is, and will continue to be, a huge mistake. For example, when looking at the United States alone, there are approximately 135 million pet dogs and cats. This is a major public health concern because, besides being neglected from ASPs, this entire population of pet animals also receive the same antimicrobials for treatment as humans. As a result, without the concurrent controlled use of antimicrobials in animal species, there will be a continued battle of antimicrobial resistance. Additionally, the CDC reports that over 75% of the emerging infectious diseases in humans are zoonotic, meaning that they originate in animal species and are then directly spread to humans through physical contact or through consumption of the animal (32). It is paramount that antimicrobial therapies are used appropriately in all species in order to truly combat the effect of AMR in the world today. In conclusion, focused research on veterinary ASPs is required in order to gain a true global representation of current AMR.

The success of ASPs is dependent on involvement and coordination of all major stakeholders, effective oversight and communication systems, dedicated sources of funding for the program, education improvement of prescribers, and finally the willingness/attitude of prescribers to follow guidelines of treatment. While some studies show that ASPs are effective methods to decrease the use of unnecessary antimicrobials and decrease the cost of patient care all while improving treatment guidelines and patient outcomes, there is still little definitive evidence that ASPs are successful at preventing or slowing down rates of AMR.

References

(1) World Health Organization. World health statistics 2022: Monitoring health for the SDGs, sustainable development goals. Geneva: WHO, 2022. URL: https://www.who.int/publications/i/item/9789240051157.

(2) Centers for Disease Control and Prevention. Antibiotic resistance threats in the United States. Atlanta, GA: CDC, 2019. URL: https://www.cdc.gov/drugresistance/pdf/threats-report/2019-ar-threats-report-508.pdf.

(3) Hayes JF. Fighting back against antimicrobial resistance with comprehensive policy and education: A narrative review. Antibiotics 2022;11(5):644.

(4) Shirazi OU, Rahman NSA, Zin CS. An overview of the hospitals' antimicrobial stewardship programs implemented to improve antibiotics' utilization, cost and resistance patterns. J Pharmacy 2022;2(1):16-30.

(5) Schmid S, Schlosser S, Gülow K, Pavel V, Müller M, Kratzer A. Interprofessional collaboration between ICU physicians, staff nurses, and hospital pharmacists optimizes antimicrobial treatment and improves quality of care and economic outcome. Antibiotics 2022;11(3):381.

(6) Alghamdi S, Berrou I, Bajnaid E, Aslanpour Z, Haseeb A, Hammad MA, et al. Antimicrobial stewardship program implementation in a Saudi medical city: An exploratory case study. Antibiotics 2021;10(3):280.

(7) Verliat F, Hemonic A, Chouet S, Le Coz P, Liber M, Jouy E, et al. An efficient cephalosporin stewardship programme in French swine production. Vet Med Sci 2021;7(2):432-9.

(8) Huong VTL, Ngan TTD, Thao HP, Tu NTC, Quan TA, Nadjm B, et al. Improving antimicrobial use through antimicrobial stewardship in a lower-middle income setting: A mixed-methods study in a network of acute-care hospitals in Viet Nam. J Glob Antimicrob Resist 2021;27:212-21.

(9) Labricciosa FM, Sartelli M, Correia S, Abbo LM, Severo M, Ansaloni L, et al. Emergency surgeons' perceptions and attitudes towards antibiotic prescribing and resistance: A worldwide cross-sectional survey. World J Emerg Surg 2018;13(1):27.

(10) Moriyama Y, Ishikane M, Kusama Y, Matsunaga N, Tajima T, Hayakawa K, et al. Nationwide cross-sectional study of antimicrobial stewardship and antifungal stewardship programs in inpatient settings in Japan. BMC Infect Dis 2021;21(1):355.

(11) Pallares C, Hernández-Gómez C, Appel TM, Escandón K, Reyes S, Salcedo S, et al. Impact of antimicrobial stewardship programs on antibiotic consumption and antimicrobial resistance in four Colombian healthcare institutions. BMC Infect Dis 2022;22(1):420.

(12) Kim SH, Yoon JG, Park HJ, Won H, Ryoo SS, Choi E, et al. Effects of a comprehensive antimicrobial stewardship program in a surgical intensive care unit. Int J Infect Dis 2021;108:237-43.

(13) Veepanattu P, Singh S, Mendelson M, Nampoothiri V, Edathadatil F, Surendran S, et al. Building resilient and responsive research collaborations to tackle antimicrobial resistance—Lessons learnt from India, South Africa, and UK. Int J Infect Dis 2020;100:278-82.

(14) Sarwar MR, Saqib A, Iftikhar S, Sadiq T. Knowledge of community pharmacists about antibiotics, and their perceptions and practices regarding antimicrobial stewardship: A cross-sectional study in Punjab, Pakistan. Infect Drug Resist 2018;11:133-45.

(15) Firouzabadi D, Mahmoudi L. Knowledge, attitude, and practice of health care workers towards antibiotic resistance and antimicrobial stewardship programmes: A cross-sectional study. J Eval Clin Pract 2020;26(1):190-6.
(16) Nasr ZG, Alhaj Moustafa D, Dahmani S, Wilby KJ. Investigating pharmacy students' therapeutic decision-making with respect to antimicrobial stewardship cases. BMC Med Educ 2022;22(1):511.
(17) Sundvall PD, Skoglund I, Hess-Wargbaner M, Åhrén C. Rational antibiotic prescribing in primary care: Qualitative study of opportunities and obstacles. BJGP Open 2020;4(4):bjgpopen20X101079.
(18) McIsaac W, Kukan S, Huszti E, Szadkowski L, O'Neill B, Virani S, et al. A pragmatic randomized trial of a primary care antimicrobial stewardship intervention in Ontario, Canada. BMC Fam Pract 2021;22(1):185.
(19) Kishida N, Nishiura H. Accelerating reductions in antimicrobial resistance: Evaluating the effectiveness of an intervention program implemented by an infectious disease consultant. Int J Infect Dis 2020;93:175-81.
(20) Adegbite BR, Edoa JR, Schaumburg F, Alabi AS, Adegnika AA, Grobusch MP. Knowledge and perception on antimicrobial resistance and antibiotics prescribing attitude among physicians and nurses in Lambaréné region, Gabon: A call for setting-up an antimicrobial stewardship program. Antimicrob Resist Infect Control 2022;11(1):44.
(21) Díaz-Madriz JP, Cordero-García E, Chaverri-Fernández JM, Zavaleta-Monestel E, Murillo-Cubero J, Piedra-Navarro H, et al. Impact of a pharmacist-driven antimicrobial stewardship program in a private hospital in Costa Rica. Rev Panam Salud Pública 2020;44:e57.
(22) da Silva RMR, de Mendonça SCB, Leão IN, dos Santos QN, Batista AM, Melo MS, et al. Use of monitoring indicators in hospital management of antimicrobials. BMC Infect Dis 2021;21(1):827.
(23) Pauwels I, Versporten A, Vermeulen H, Vlieghe E, Goossens H. Assessing the impact of the Global Point Prevalence Survey of Antimicrobial Consumption and Resistance (Global-PPS) on hospital antimicrobial stewardship programmes: Results of a worldwide survey. Antimicrob Resist Infect Control 2021;10(1):138.
(24) Mahmoudi L, Sepasian A, Firouzabadi D, Akbari A. The impact of an antibiotic stewardship program on the consumption of specific antimicrobials and their cost burden: A hospital-wide intervention. Risk Manag Healthc Policy 2020;13:1701-9.
(25) Shin JH, Mizuno S, Okuno T, Itoshima H, Sasaki N, Kunisawa S, et al. Nationwide multicenter questionnaire surveys on countermeasures against antimicrobial resistance and infections in hospitals. BMC Infect Dis 2021;21(1):234.
(26) Tartof SY, Chen LH, Tian Y, Wei R, Im T, Yu K, et al. Do inpatient antimicrobial stewardship programs help us in the battle against antimicrobial resistance? Clin Infect Dis 2021;73(11):e4454-62.
(27) Centers for Disease Control and Prevention. COVID-19: U.S. impact on antimicrobial resistance, special report 2022. Atlanta, GA: CDC, 2022. URL: https://stacks.cdc.gov/view/cdc/117915.
(28) Tomczyk S, Taylor A, Brown A, de Kraker MEA, El-Saed A, Alshamrani M, et al. Impact of the COVID-19 pandemic on the surveillance, prevention and control of

antimicrobial resistance: A global survey. J Antimicrob Chemother 2021;76(11):3045-58.

(29) Saleem Z, Hassali MA, Hashmi FK, Godman B, Ahmed Z. Snapshot of antimicrobial stewardship programs in the hospitals of Pakistan: Findings and implications. Heliyon 2019;5(7):e02159.

(30) Bolten BC, Bradford JL, White BN, Heath GW, Sizemore JM, White CE. Effects of an automatic discontinuation of antibiotics policy: A novel approach to antimicrobial stewardship. Am J Health-Syst Pharm 2019;76(3):S85-90.

(31) Lubwama M, Onyuka J, Ayazika KT, Ssetaba LJ, Siboko J, Daniel O, et al. Knowledge, attitudes, and perceptions about antibiotic use and antimicrobial resistance among final year undergraduate medical and pharmacy students at three universities in East Africa. PLoS One 2021;16(5):e0251301.

(32) Centers for Disease Control and Prevention. Zoonotic diseases. Atlanta, GA: CDC, 2021. URL: https://www.cdc.gov/onehealth/basics/zoonotic-diseases.html.

Appendix A

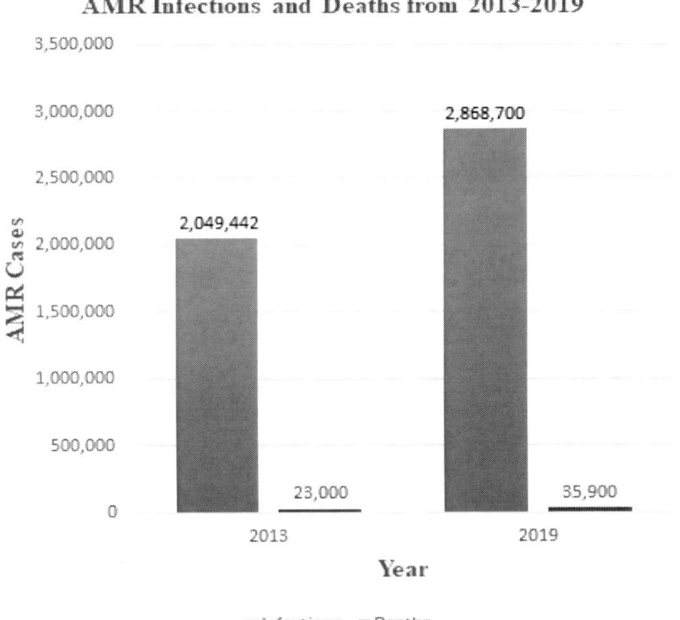

Note. The number of estimated AMR infections and deaths reported by the CDC increased by 819,258 and 12,900 respectively from 2013 to 2019.

Figure 2. AMR infections and deaths in the United States.
Comparison of the CDC's 2013 & 2019 AR Threats Reports.

Appendix B

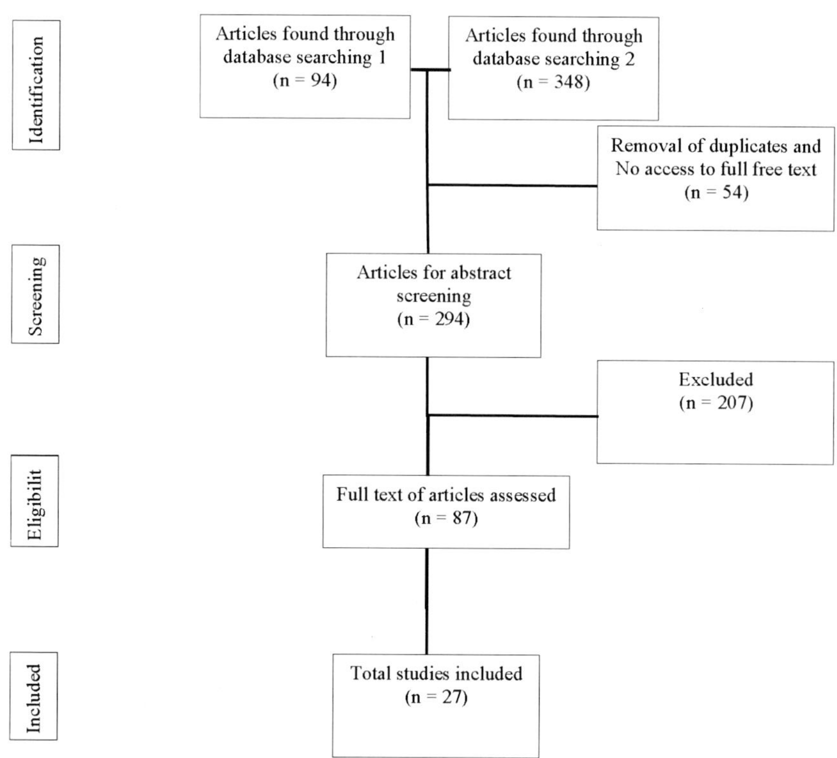

Figure 3. Study selection flowchart.

Chapter 9

A comparative review between USDA and FAO meat inspection standards and foodborne illness burdens in the United States and Grenada

Ashlyn Brewster, DVM, MPH
Sara Elzibak, BSc
and Satesh Bidaisee*, DVM, MSPH, EdD
Public Health and Preventive Medicine, St George's University, Grenada, West Indies

Abstract

The United States (US) and the Caribbean country of Grenada each have their own governing body of food safety and inspection. The US follows the Food Safety and Inspection Service (FSIS) branch of the United States Department of Agriculture (USDA), while Grenada follows the Food and Agriculture Organization of the United Nations (FAO). When foodborne illness outbreaks occur, it often can be traced back to gaps or oversights within food inspection before that food reaches the public. Therefore, this study is a narrative review and aims to compare the safety standards of USDA and FAO and any possible correlation with foodborne illness burdens using the United States and the small Caribbean Island of Grenada as representative countries. For this study, a computer search was conducted using public databases Google Scholar, PubMed, ScienceDirect, and AGORA. The US is frequently updating legislation as new research emerges while Grenada's legislative acts have

* ***Correspondence:*** Satesh Bidaisee, Professor, Public Health and Preventive Medicine, St George's University, Grenada, West Indies. Email: sbidaisee@sgu.edu

In: Public Health: Intersection of Health, Humans, Animals and the Environment
Editors: Satesh Bidaisee and Joav Merrick
ISBN: 979-8-89530-443-3

stayed unrevised for almost decades. The FAO does have similar standards regarding food safety as the US does, but they are not as enforced in Grenada and other smaller developing countries. The US is constantly tracking and surveying foodborne illness outbreaks and implementing policies in the wake of upward trends while Grenada does not have nearly as much outbreak surveillance to track and respond to trends. Grenada's inspection regulations and food safety protocols have made little improvement over the years, but much more research and many more changes are still to be done.

Introduction

When it comes to food safety and food hygiene, meat inspection protocols within slaughterhouses are not standard around the world. In the United States, inspection is under the jurisdiction of the Food Safety and Inspection Service (FSIS) within the United States Department of Agriculture (USDA). Any meat that is to be sold commercially is required to be inspected by a federal inspector within FSIS. The goal is to ensure that only food that is "safe, wholesome, and properly labeled" is allowed to be released into the food chain (1). Livestock, defined in the Federal Meat Inspection Act as cattle, sheep, swine, and goat, must be inspected throughout the entire slaughtering process, and the whole carcass, parts of it, or any product intended for human consumption must always be federally inspected. Inspection personnel should also ensure humane slaughtering of every animal (2). While there are several exemptions to these rules, this study will not review these.

In Grenada, as well as other small and developing countries of the Caribbean, meat product import exceeds meat export. As of 2019, Grenada slaughters mostly sheep and goats followed by poultry and few cattle (3). The Veterinary and Livestock Division of the country's Ministry of Agriculture and Lands is currently in charge of "treatment of animals, risk assessment prior to meat import, ante-mortem inspection…prior to slaughter, quarantine, disease surveillance, [and] issuing permits for live animals, meat, and meat products" (4). Inspections within "production, processing, packing, packaging, storing, transporting, distributing, and selling food products" are regularly performed in order to ensure the national health standards are met (4). Implementation of the Hazard Analysis and Critical Control Point (HACCP) system, and other quality assurance programs are required in Grenada slaughterhouse plants, especially for export products. Caribbean countries are also required to have an "analytical laboratory that meets international quality standards" (5).

If there are holes or weaknesses within the quality assurance and inspection protocols for meat and food products in any country, potential outbreaks of foodborne illnesses can occur. Foodborne illnesses are a prominent public health issue since they not only cause outbreaks within the population and make people severely sick, but can also become an economic burden, especially in lesser developed countries (6). The Centers for Disease Control (CDC) estimates in the US that 48 million people a year get sick and about 3,000 people die from foodborne illnesses (7). Literature shows that there is little evidence on various foodborne illnesses in low-income countries such as Grenada due to less surveillance on the subject (8). Indar et al., (9) reported that despite limited surveillance data, there is still an increasing trend in foodborne illness outbreaks from *salmonella, shigella* and *campylobacter* (9). Surveillance of both foodborne illnesses and animal diseases is vital in preventing these outbreaks from occurring in the first place (10).

Risk assessments have been performed within Caribbean countries, concluding that the number one concern would be to avoid contamination from animals used within the food supply (11). Therefore, this study aims to identify any differences between the standards in inspection and quality assessment of meat products between the USDA in the US and FAO in Grenada, and if the differences in standards have an effect or influence on foodborne illness burdens within the two countries.

If a significant difference is found, this study suggests an exploration of possible solutions for improving food hygiene. When offering solutions for improvement, it would be vital to remember that Grenada is a developing country, and resources are limited. It is of importance to perform a cost analysis of implementing safety protocols before moving forward with application (12). One Health puts the focus on an all-encompassing relationship between people, animals, and the environment, and using agriculture to raise animal food products while preventing threats to the population via food contamination is key (13). It would then be impertinent that solutions should take all of this into account as promotion and awareness of the importance of One Health may help decrease outbreaks as well as financial burden (14).

Our study

Multiple databases were used to perform the literature search. These were Google Scholar, PubMed, ScienceDirect, and AGORA. The following Boolean search terms were used:

1. Food safety OR food hygiene
2. Slaughterhouse OR abattoir AND protocol* (OR regulations)
3. USDA AND FAO
4. Foodborne illness (OR disease) AND USA OR Grenada (OR Caribbean OR West Indies)
5. Various combinations of the above searches were used

Articles were limited to papers that were in English, published in the last 10 years, and were available as full free texts. It was important for articles to be no more than 10 years old because older papers would likely be out of date and an inaccurate reflection of more recent data and trends currently happening. The type and prevalence of foodborne illness outbreaks have changed over the years, so a review of the most recent trends in both the US and Grenada was conducted. An exception to limiting articles to the last 10 years included older resources that included policies and regulations that have not been updated and, therefore, are still relevant.

Titles and abstracts of papers were manually read and screened for relevance, using guidelines from the Preferred Reporting Items for Systematic reviews and Meta-Analyses (PRISMA) (15). Resources were limited to topics based on meat products from poultry and livestock species of cattle, sheep, and goats; horses were excluded. The US was used as a whole, and no specific state was excluded in this review. Since Grenada is a small country, there was limited data on specific topics, and therefore, the search was extended to the Caribbean as a whole, including other similar developing countries-such as Barbados, the Bahamas, West Africa, and other countries-under FAO, but still maintaining a focus on Grenada. Disease pathogen information was limited to only those that caused common outbreaks and were food related. The reference lists of the selected articles were also used to retrieve other relevant sources. Literature selected were a total of 10 full-text articles (see Figure 1).

To analyze the sources, a qualitative data analysis was used. Since this study is a comparative analysis, literature was first organized and sorted into separate folders on Mendeley grouped by topic. These topics included food safety in Grenada, food safety in the US, foodborne diseases in Grenada, and foodborne diseases in the US. Highlighted relevant information and margin notes from the articles were made to detect comparisons. This included any mention of current legislation, descriptions of current food inspection protocols, statistics of outbreak numbers, types of foodborne illnesses, and examples of foodborne illness surveillance. Information and data from the articles were compared to each other depending on their specific topics. This meant food safety (current policies, regulations, inspection requirements) in the US was compared to food safety in Grenada and foodborne illnesses (pathogens, trends, outbreaks) in the US were also compared to foodborne illnesses in Grenada.

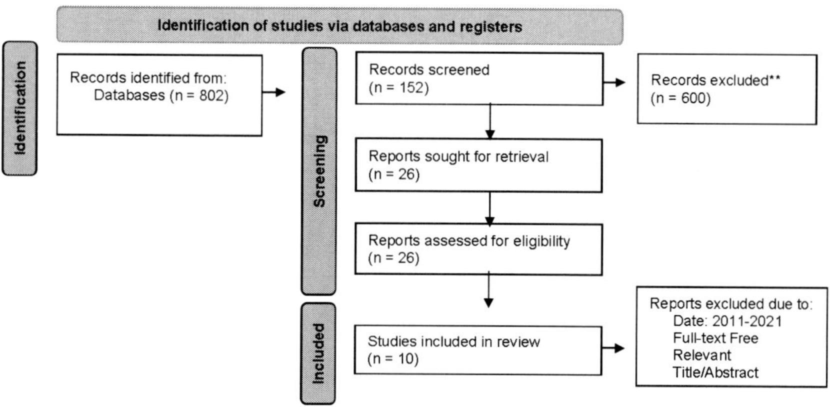

Figure 1. Literature selection process using guidelines from PRISMA.

Discussion

Meat inspection in the United States is extremely strict and follows detailed protocols. The 1906 Federal Meat Inspection Act and Pure Food and Drugs Act are essentially the foundation for the US' legislation regarding safe meat; the passing of the 1967 Wholesome Meat Act expanded the meat inspection program standards; and the FDA Food Safety Guidance Documents provided accessible and specific instructions on how to safely handle various food products (16). The FDA's Food Safety Modernization Act (FSMA) essentially calls for more frequent facility inspections, access to records, demand for recalls, detailed development of facility HACCP plans, as well as the potential for accreditation for meat testing labs (17). Facilities are required to have mandated inspection with frequency based on risk assessment, to keep documentation of records that are accessible by the FDA, and to have certain food testing done by accredited laboratories only. The FDA has an extensive process of administrative actions for facilities that do not meet requirements, which range from issuing warning letters to refusing imports to disqualifying and suspending facilities from processing food and even to court orders and criminal prosecutions (18).

Currently, the health sector of Grenada is the Ministry of Health (MOH), which collaborates with other governing bodies such as the Caribbean Public Health Agency (CARPHA) and the Veterinary and Livestock Unit of the Ministry of Agriculture (MOA) with regards to public health. The Public Health Act was enacted in 1958, but it serves as an outdated regulation because it has not been updated since. The Food and Drugs Act of 1986 was revised back in 2011, but still has gaps in what constitutes proper food safety expectations (19). The Food and Drugs Act clearly outlines that a person who is found selling food that contains harmful substances, is unfit for human consumption, rotten, or stored improperly is guilty of an offence and that the Ministry can fine, seize, and send the guilty party to court. Grenada is often does not enforce facility inspections and food safety violations like the US does. This can be attributed to the limited capacity of staffing and resources, as evidence shows none of the legislative acts are properly enforced (20).

Back in the late 20th century, the US saw an uprising in agriculture technology as well as an increased demand for food production (21). There is a multitude of research that has led the US to understand how processed food affects human health. The only meat or meat product allowed to be served in retail stores and restaurants is meat that has been inspected by a licensed public health inspector from pre-slaughter, during slaughter, and all the way through to processing (2). Not only are animals inspected, but facilities and equipment are as well. The use of labels has also strengthened the standards of the US. These labels include warnings for raw consumption, for example, and certified official inspected stamps that say the meat is safe to eat (2).

Grenada does not offer those same strict standards as the US does, even though it is underneath the FAO. The World Organization for Animal Health, also known as the Office International des Epizooties (OIE) collaborated with the World Health Organization and established a set of guidelines that focus on the improvement of the health and welfare of animals used for production. These guidelines, referred to as The Terrestrial Animal Health Code (or the Terrestrial Code), are "based on the most recent scientific and technical information…to protect animal health and welfare and veterinary public health during production and trade in animals and animal products, and in the use of animals" (22). As of 2017, there are 180 Member Countries with Grenada having observer status (20). A study done by St George's University in the Department of Public Health and Preventative Medicine in 2016 reviewed how compliant Grenada's poultry industry was with regards to the guidelines of the OIE's Terrestrial Code. Results illustrated that Grenada was either "non- or partially compliant with guidelines on animal disease diagnosis, surveillance and notification, risk analysis, and veterinary public health" (20). The MOA is in charge of performing both ante- and post-mortem inspections on animals going to slaughter. However, evidence shows that animals unfit for human consumption often slip into the local food chain because of a lack of MOA licensed personnel availability and a lack of regular "microbiological, serological, or other testing," of meat (20). On a positive note, Grenada's abattoirs have made improvements in the right direction with more training classes for workers, donations of sanitizing equipment and materials, and implementation of HACCP protocols, but there are still many gaps in Grenada's legislation and government that need to be rectified in order to strengthen food supply security and food safety.

A study done by the CDC concluded that the majority of foodborne illnesses within the US do not actually come from meat, but from produce (7). This can likely be attributed to the high standards in meat inspection set by the USDA (23). Regardless, any source of foodborne illness outbreaks has the ability to be detrimental to any economy, even in developed countries. Economically, foodborne illness outbreaks are huge burdens and improving food hygiene can prevent loss of funds, which can be put towards strengthening healthy and substantial food security (24). In the US, the USDA, FDA, and even the Environmental Protection Agency (EPA) are not only allowed to enact legislation but are also "legal enforcers to protect public health and safety" (24).

Surveillance of foodborne illness outbreaks in the US is carried out by various agencies. There is a list of reportable diseases that each state must adhere to. Clinicians and laboratories are legally required to report diagnoses of these reportable diseases to their local health departments, which in turn report to the CDC (16). The CDC's methods for detecting outbreaks include molecular testing, genome sequencing, quantitative microbial risk assessments, and other technologies (16). Though the surveillance of foodborne illnesses in the US is strong today, there is still always room for improvement and continued research.

While foodborne illness outbreaks greatly impact any country, it is evident that the impact is greater in developing countries as they just do not have resources to successfully combat, monitor, and control outbreaks (24). As previously mentioned, because food safety standards are not as rigid in Grenada, it is more likely for unsafe food to get into local markets. The risk is especially high as it is popular for Caribbean street vendors to sell "ready to eat foods" that may expose the public to contacting various pathogens and bacteria (25). Another potential impact worth noting has to do with tourism. Grenada relies heavily on tourism from cruise ships and other travelers to support the country's economy and Gross Domestic Product (GDP) (25). When tourists and workers are exposed to and get sick with foodborne pathogens, outbreaks can negatively affect tourism by causing increased strain on an already weak healthcare system as well as future avoidance of returning tourists (25).

It is estimated that less than ten percent of foodborne illness cases are reported whereas less than one percent of cases are reported in developing nations (24). In Grenada, epidemiologic surveillance is the responsibility of the MOH, while zoonotic disease documentation is the responsibility of the MOA (20). The issue is that there are discrepancies in the roles that the MOA plays in reporting, which include irregular reporting meetings, verbal debriefings rather than written official statements, and a lack of research "directly focused on zoonotic diseases etiology and trends" in Grenada (20). While the MOH reports findings to the CARPHA, who then report to WHO and so on, studies found that MOA is not included in Grenada's national surveillance network (20). Disease surveillance requires teamwork and collaboration between various stakeholders in order to contribute to an all-encompassing One Health perspective, which would help strengthen a community's response to zoonoses, such as foodborne pathogens.

Conclusion

The US is not perfect and the USDA's food safety regulations do not prevent 100% of foodborne illness outbreaks, but the high standards that slaughter plants are held to have improved by leaps and bounds over decades. The extensive research and revision of legislation have led to the US having a much safer public meat supply. Zoonoses are not eradicated, but technology is advancing so that outbreaks are more easily contained. Grenada's slaughter plant regulations have also made improvements over the years, but there is still a long way to go. It would be beneficial for Grenada to revisit their regulations and make amends that accurately reflect contemporary changes regarding public health and food safety. Recommendations to Grenada should be to first work on improving surveillance of foodborne disease outbreaks so that trends can be tracked. A call to action of improved collaboration between governing bodies, both human and veterinary related, in order to reach to a One Health outlook could potentially help Grenada prevent economic burdens due to the gaps in food safety and foodborne illness outbreaks. The hiring of more MOA licensed personnel to perform regulatory animal and facility inspections could also be improved upon so that outbreaks can be traced back to the carcass, animal, and farm before any outbreak gets worse. Hiring more personnel may also help increase enforcement of laws. Lastly, continued research should be conducted within the country of Grenada regarding epidemiologically significant pathogens related to food safety, strengthening standards within food processing plants, as well as providing population education on the relevant topics of food handling, food safety, and zoonotic diseases.

References

(1) Federal Meat Inspection Act 1906 (FSIS).

(2) Food Safety Inspection Service. Summary of federal inspection requirements for meat products. USDA 2015. URL: https://www.fsis.usda.gov/sites/default/files/media_file/2021-02/Fed-Food-Inspect-Requirements.pdf#:~:text=The%20Federal%20Meat%20Inspection%20Act%20%28FMIA%29%20requires%20that,must%20be%20inspected%20and%20passed%20for%20human%20consumption.

(3) Food and Agriculture Organization of the United Nations. FAOSTAT: Grenada. FAO. URL: http://www.fao.org/faostat/en/#country/86.

(4) The Grenada Food and Nutrition Council. The food and nutrition policy and plan of action for Grenada. PAHO 2007. URL: https://extranet.who.int/nutrition/gina/sites/default/filesstore/GRD%202007%20Food%20and%20nutrition%20policy.pdf.

(5) Food and Agriculture Organization of the United Nations. National food safety systems in the Americas and the Caribbean - A situation analysis. FAO. URL: http://www.fao.org/3/a0394e/A0394E20.htm.

(6) Guerra MM, de Almeida AM, Willingham AL. An overview of food safety and bacterial foodborne zoonoses in food production animals in the Caribbean region. Trop Anim Health Prod 2016;48(6):1095-108.

(7) Centers for Disease Control and Prevention. Foodborne germs and illnesses. CDC 2013. URL: https://www.cdc.gov/foodsafety/foodborne-germs.html.

(8) Grace D. Food safety in low and middle income countries. Int J Environ Res Public Health 2015;12(9):10490-507.

(9) Indar-Harrinauth L, Daniels N, Prabhakar P, Brown C, Baccus-Taylor G, Comissiong E, et al. Emergence of salmonella enteritidis phage type 4 in the Caribbean: Case-control study in Trinidad and Tobago, West Indies. Clin Infect Dis 2001;32(6), 890-6.
(10) Vidal SM, Fajardo PI, Gonzalez CG. Veterinary education in the area of food safety (including animal health, food pathogens and surveillance of foodborne diseases). Rev Sci Tech 2013;32(2):425-31.
(11) Percedo Abreu MI, Guitián J, Herbert-Hackshaw K, Pradel J, Bournez L, Petit-Sinturel M, et al. Developing a disease prevention strategy in the Caribbean: The importance of assessing animal health-related risks at regional level. Rev Sci Tech 2011;30(3):725-31.
(12) Viator CL, Muth MK, Brophy JE, Noyes G. Costs of food safety investments in the meat and poultry slaughter industries. J Food Sci 2017;82(2):260-9.
(13) Gebreyes WA, Dupouy-Camet J, Newport MJ, Oliveira CJ, Schlesinger LS, Saif YM, et al. The global one health paradigm: Challenges and opportunities for tackling infectious diseases at the human, animal, and environment interface in low-resource settings. PLoS Negl Trop Dis 2014;8(11):e3257.
(14) Rist CL, Arriola CS, Rubin C. Prioritizing zoonoses: A proposed one health tool for collaborative decision-making. PLoS ONE 2014;9(10):e109986.
(15) Page MJ, McKenzie JE, Bossuyt PM, Boutron I, Hoffmann TC, Mulrow CD, et al. The PRISMA 2020 statement: An updated guideline for reporting systematic reviews. BMJ 2021;372:n71.
(16) Doyle MP, Erickson MC, Alali W, Cannon J, Deng X, Ortega Y, et al. The food industry's current and future role in preventing microbial foodborne illness within the United States. Clin Infect Dis 2015;61(2):252-9.
(17) Food Safety Modernization Act 2011 (FDA).
(18) Food and Drug Administration. Compliance & Enforcement. FDA 2022. URL: https://www.fda.gov/animal-veterinary/compliance-enforcement#enforcement .
(19) Pan American Health Organization. Grenada. Geneva: WHO, 2022. URL: https://hia.paho.org/en/countries-22/grenada-country-profile.
(20) Glasgow L, Forde M, Fletcher S, Keku E. Compliance with the world organisation for animal health guidelines for poultry production in Grenada. World Poultry Sci J 2017;73(3):515-26.
(21) Wu F, Rodricks JV. Forty years of food safety risk assessment: A history and analysis. Risk Anal 2020;40(S1):2218-30.
(22) World Organisation for Animal Health. Terrestrial Animal Health Code. OIE 2021. URL: https://www.oie.int/en/what-we-do/standards/codes-and-manuals/terrestrial-code-online-access/.
(23) Painter JA, Hoekstra RM, Ayers T, Tauxe RV, Braden CR, Angulo FJ. Attribution of foodborne illnesses, hospitalizations, and deaths to food commodities by using outbreak data, United States, 1998–2008. Emerg Infect Dis 2013;19(3):407-15.
(24) Fung F, Wang HS, Menon S. Food safety in the 21st century. Biomed J 2018;41(2):88-95.
(25) Lee B. Foodborne disease and the need for greater foodborne disease surveillance in the Caribbean. Vet Sci 2017;4(3):40.

Chapter 10

Zoonotic microorganisms as agents of agroterrorism in the United States

Sara Schectman, DVM, MPH
Sara Elzibak, BSc
and Satesh Bidaisee*, DVM, MSPH, EdD
Public Health and Preventive Medicine, St George's University, Grenada, West Indies

Abstract

Zoonotic diseases, germs spread between animals and people, are common worldwide as scientists estimate more than 6 out of every 10 infectious diseases are spread from animals. The deliberate release of zoonotic agents for agroterrorism can cause illness, death and large-scale economic loss. The purpose of this narrative review is to further comprehend the potential for deliberate misuse of two commonly known zoonotic microorganisms, *brucella* and *bacillus anthracis,* and the ways in which the United States has addressed vulnerabilities and established defensive strategy. Although most agents are virtually eradicated in the United States livestock population, majority of surrounding countries do not have eradication or vaccination protocols which poses a threat to the global food trade. Literature regarding microorganisms was collected and analyzed using advanced search engines, and inclusion and exclusion criteria. Quality assessments of the literature were conducted in order to extract the most relevant information regarding past, present, and future direction of preparation and preparedness strategies. Continued collaboration amongst veterinarians, physicians, and federal agencies

---

* **Correspondence:** Satesh Bidaisee, Professor, Public Health and Preventive Medicine, St George's University, Grenada, West Indies. Email: sbidaisee@sgu.edu

In: Public Health: Intersection of Health, Humans, Animals and the Environment
Editors: Satesh Bidaisee and Joav Merrick
ISBN: 979-8-89530-443-3

remains crucial in national security efforts and in maintaining public confidence in the security of the food supply.

Introduction

Zoonotic diseases are caused by viruses, bacteria, parasites and fungi that are transmitted between vertebrate animals and man (1-4). Dating as far back as the 6th century BCE, zoonotic microorganisms have been weaponized against humans, animals, and agriculture (5). Agroterrorism is of heightened concern as deliberate attacks threaten the food supply and economic stability (6). The agriculture industry is one of the largest sectors in the United States (US) and in 2020, contributed almost 20% to the GDP while comprising over 10% of exports (7). The FDA estimates the cost of an attack on the US food supply could total over $100 billion (8).

Table 1. CDC bioterrorism agents and eradication status in the United States

	Agents/Diseases	Eradication status in agriculture
Category A	Anthrax, Botulism, Plague, Smallpox, Tularemia, Viral hemorrhagic fever	
Category B	Brucellosis, Epsilon toxin, Salmonella, Escherichia coli, Shigella, Glanders, Melioidosis, Psittacosis, Q fever, Ricin toxin, Staphylococcal enterotoxin B, Typhus fever, Viral encephalitis, Vibrio cholerae, Cryptosporidium parvum	Brucellosis

The CDC (Centers for Disease Control and Prevention) has prioritized bioterrorism agents into three respective categories: A, B and C (see Table 1). *Bacillus anthracis*, a group A agent, and the etiological agent of anthrax is a soil-borne bacteria that occurs naturally in the environment worldwide and infects livestock through ingestion of contaminated water, soil, or vegetation (9). According to a meta-analysis compiled in 2020, the prevalence of anthrax in livestock in North America was 21% (10). Anthrax regained the attention of the general public in 2001 due to the anthrax letters terrorist attacks. Shortly after the attacks, congress signed into law the Bioterrorism Preparedness and Response Act of 2002, designed to protect the nation from bioterrorism and the food supply against intentional contamination. Furthermore, the FDA Food Safety Modernization Act was enacted in 2011 with a focus on foodborne illness prevention.

Brucella, a group B agent and the etiological agent of brucellosis, is another highly contagious zoonotic bacteria with a low infectious dose that devastates livestock and wildlife worldwide (11). Brucellosis can be easily transmitted through aerosols, unpasteurized milk, undercooked meat, or secretions (12). Malicious introduction can result in loss of production, trade consequences, and compromised food security. In 1946, 124,000 herds were affected by brucella totaling over $400 million in annual loss (13). In 1954, the National Brucellosis Eradication Program was established as a cooperative effort between the federal government, states, and livestock producers. This eradication program consists of vaccination, testing, quarantine, and slaughter. By the early 2000's, national eradication in cattle was mostly achieved and, as of April 2022, all states were classified as free from brucellosis.

Our review

In this narrative review, clinical data regarding agents of agroterrorism were researched and reviewed using various search strategies and online databases. First, the Center for Disease Control and Prevention website was utilized in order to determine a list of the zoonotic biowarfare agents and their respective categories (A, B or C). Inclusion criteria were determined to include category A and B agents, agroterrorism agents, agents affecting livestock and agents affecting plants. Exclusion criteria included category C agents, agents that were not present in the US and agents that were not zoonotic. Further research was conducted for category A and B zoonotic biowarfare agents using the St George's University e-library, Google Scholar, Science Direct, PubMed and EBSCO. Key words for searches included: agroterrorism, biowarfare, livestock, zoonotic, brucella, brucellosis, undulant fever, Malta fever, Mediterranean fever, bacillus anthracis, anthrax. Initially, 41 relevant and accessible peer-reviewed articles were selected. Abstracts were read to determine which biological agents were discussed. Further research regarding the past and present use and threat of these microorganisms in the US was conducted. No date restrictions were utilized. Ultimately, 20 full articles were accessed to compile this narrative review. Ethical approval was not required.

Discussion

Bacillus anthracis and *Brucella* are found to be popular topics in published literature pertaining to agroterrorism. Based on a review of 20 articles, anthrax and brucellosis are not as common in US livestock or human incidences. However, the fatality rate associated with a potential outbreak of either zoonotic disease could be devastating for both agriculture and mankind. Ultimately, it was found that anthrax and brucellosis are more commonly found in developing countries and countries without public health programs. For example, brucellosis is endemic in India in both humans and animals due to disregard for PPE, poor vaccination programs, poor biosecurity measures and lack of public education (14). In the United States, brucellosis is endemic only in the Greater Yellowstone Area among bison and elk populations. According to a study, about 60% of bison and 40% of elk in the GYA are seropositive for *brucella* (15). Management strategies to reduce seroprevalence include test and slaughter, vaccination, and low-density feeding (16). According to a 2018 study using the Maxent model, it was concluded that temperature, soil composition, and livestock density influence prevalence of anthrax spores in the environment worldwide (17). With that, anthrax is found to still be enzootic in parts of Africa, Central Asia, and the Middle East due to the ecosystems which promote the survival of anthrax spores (18). Other reasons that anthrax and brucella are more commonly found in these developing countries include lack of public health programs, lack of biosecurity and overall lack of public education.

The United States has, over the decades, made great advances in law and technology that has allowed for increased national security. However, limitations to security include the increasing volumes of agricultural imports into the United States. From 2019 to 2020, US imports of agricultural products increased 2.2% (19). The top suppliers of imported agricultural products into the US continue to be Mexico, Canada, and France. It would be advantageous for future studies on the threat of agroterrorism to focus on specific products imported as well as the laws and security measures of those countries who largely contribute imports.

Conclusion

In summary, this narrative review aimed to provide collective insight on two known agroterrorism agents, *bacillus antrhacis* and *brucella,* and the impact these two agents have made in the past, present, and future regarding direction of national food and livestock security. Various programs and laws have been enacted over the decades in order to improve detection, preparedness and prevention. Ultimately, it can be concluded that early detection systems such as cyber security, syndromic surveillance, and DIVA (differentiating infected from vaccinated animals) vaccines are being utilized and consistently revised (20). It is critical for government sectors, public health officials, farmers, and veterinarians to continue to collaborate and address vulnerabilities in order to continue to secure the nation's agriculture and food supply in the fight against agroterrorism.

References

(1) Centers for Disease Control and Prevention. Zoonotic Diseases. One Health 2021. URL: https://www.cdc.gov/onehealth/basics/zoonotic-diseases.html.

(2) Monov A, Dishovsky C. Medical aspects of chemical and biological terrorism: Biological terrorism and traumatism. Bulgaria: Military Academy, 2004.

(3) Keremidis H, Appel B, Menrath A, Tomuzia K, Normark M, Roffey R, et al. Historical perspective on agroterrorism: Lessons learned from 1945 to 2012. Biosecur Bioterr 2013;11(1):S17-24.

(4) Chomel BB. Zoonoses. In: Schaechter M, ed. Encyclopedia of microbiology. New York: Academic Press, 2009:820-9.

(5) Olson D. Agroterrorism: Threats to America's economy and food supply. Law Enforcement Bulletin – Federal Bureau Investigation 2012. URL: https://leb.fbi.gov/articles/featured-articles/agroterrorism-threats-to-americas-economy-and-food-supply.

(6) Tumbarski YD. Foodborne zoonotic agents and their food bioterrorism potential: A review. Bulg J Vet Med 2020;23(2):147-59.

(7) United States Department of Agriculture. Farming and farm income. USDA 2021. URL: https://www.ers.usda.gov/data-products/ag-and-food-statistics-charting-the-essentials/farming-and-farm-income/.

(8) Schneider KR, Schneider RG, Archer DL, Webb CD, Gutierrez A. Agroterrorism in the US: An overview. UF/IFAS Dep Food Sci Human Nutr 2005. URL: https://edis.ifas.ufl.edu/publication/FS126.

(9) Goel AK. Anthrax: A disease of biowarfare and public health importance. World J Clin Cases 2015;3(1):20-33.

(10) Sushma B, Shedole S, Suresh KP, Leena G, Patil SS, Srikantha G. An estimate of global anthrax prevalence in livestock: A meta-analysis. Vet World 2021;14(5):1263-71.
(11) Godfroid J, Scholz HC, Barbier T, Nicolas C, Wattiau P, Fretin D, et al. Brucellosis at the animal/ecosystem/human interface at the beginning of the 21st century. Prev Vet Med 2011;102(2):118-31.
(12) US Government Publishing Office. Agroterrorism's perfect storm: Where human animal disease collide. US GPO 2006. URL: https://www.govinfo.gov/content/pkg/CHRG-109hhrg35566/html/CHRG-109hhrg35566.htm.
(13) Monke J. CRS Report for Congress - Agroterrorism: Threats and Preparedness. Congressional Res Ser 2007. URL: https://fas.org/sgp/crs/terror/RL32521.pdf.
(14) Tiwari HK, Proch V, Singh BB, Schemann K, Ward M, Singh J, et al. Brucellosis in India: Comparing exposure amongst veterinarians, para-veterinarians and animal handlers. One Health 2021;14:100367.
(15) Hobbs NT, Geremia C, Treanor J, Wallen R, White PJ, Hooten MB, et al. State-space modeling to support management of brucellosis in the Yellowstone bison population. Ecological Monogr 2015;85(4):525-56.
(16) Boroff K, Kauffman M, Peck D, Maichak E, Scurlock B, Schumaker B. Risk assessment and management of brucellosis in the southern greater Yellowstone Area (ii): Cost-benefit analysis of reducing elk brucellosis prevalence. Prev Vet Med 2016;134:39-48.
(17) Walsh MG, de Smalen AW, Mor SM. Climatic influence on anthrax suitability in warming northern latitudes. Sci Rep 2018;8(1):9269.
(18) Muturi M, Gachohi J, Mwatondo A, Lekolool I, Gakuya F, Bett A, et al. Recurrent anthrax outbreaks in humans, livestock, and wildlife in the same locality, Kenya, 2014–2017. Am J Trop Med Hyg 2018;99(4):833-9.
(19) US International Trade Commission. Agricultural products. US ITC 2020. URL: https://www.usitc.gov/research_and_analysis/tradeshifts/2020/agriculture.htm.
(20) Elbers A, Knutsson R. Agroterrorism targeting livestock: A review with a focus on early detection systems. Biosecur Bioterror 2013;11(1):S25-35.

Chapter 11

An ecological model: A solution to the increased risk of emerging disease linked to environmental changes impacting the global health interface stability

Janine N Wettergren, DVM, MPH
Sara Elzibak, BSc
and Satesh Bidaisee*, DVM, MSPH, EdD
Public Health and Preventive Medicine, St George's University, Grenada, West Indies

Abstract

There is a major imbalance in the global health interface which encompasses when aiming to improve the ecosystems in which all living things thrive. It is crucial to recognize the importance of reaching environmental stability and biodiversity conservation, on a local and global scale, as the crucial indicators for economic burden, ecological destruction, and existing unsustainable trends. The significance of restoring natural ecosystems and preventing further devastation to the planet's unique biomes offers the opportunity for increased potential of critical emerging disease occurrence, furthering the vulnerability of the planet's global health. The objective is to encourage the consideration of the cost and benefits of improving and sustaining crucial ecological factors as an equal contender of a mechanism for disease prevention and control. This narrative review uses an overview of 102 papers, retrieved from a computerized database, AGORA: Research4lLife. It discusses

* ***Correspondence:*** Satesh Bidaisee, Professor, Public Health and Preventive Medicine, St George's University, Grenada, West Indies. Email: sbidaisee@sgu.edu

In: Public Health: Intersection of Health, Humans, Animals and the Environment
Editors: Satesh Bidaisee and Joav Merrick
ISBN: 979-8-89530-443-3

various, prominent determinants of disease and their relationship to current environmental changes; the possible challenges associated with developing ecological stability and sustainable development; and the importance of biodiversity conservation as the proposed mechanism for reaching ecological stability. The proposed solution is prioritizing environmental stability through biodiversity conservation and sustainable development to combat the increased risk of emerging and re-emerging diseases.

Introduction

The Centers for Disease Control and Prevention (CDC) have a collaborative Emerging Infections Program that during 2010-2020 had set goals of improving health equity as a priority of their public health agenda by adding additional social and economic determinants of health with the intention of improving stratification of data to better work towards health equity (1). The biggest concern is that climate change can accelerate the process of disease emergence and transmission, making it a crucial priority of focus to prevent or limit the intensity and range across species and geographical areas of future outbreaks (2, 3).

The economic burden of disease across the globe plays a crucial role in health development through generations in high-income and low-income countries alike (4-6). Approximately 63% of the world's death is a result of non-communicable diseases, such as cardiovascular disease, cancer, respiratory diseases and diabetes, ranging from high-income to low-income countries, and young to old patients affected (4, 6, 7). There is an estimated loss of US$30 trillion (2012) to US$47 trillion (2019) due to non-communicable diseases over the next two decades, which represents about 50-75% of the global GDP in 2010 (4, 6).

Over the past few generations, the impact of the roles that different species has played in the stability of local environmental systems requires careful consideration when interventions are being discussed (8). The importance of protecting natural resources, biodiversity conservation, and the prevention of unsustainable development requires valuable environmental monitoring programs to provide valuable information on the interfaces of ecological changes (8, 9). A global multi-system approach is a priority as emerging disease risk and occurrence increase with a continued threat to global health, and devastating outcomes to the world's most vulnerable populations (9, 10).

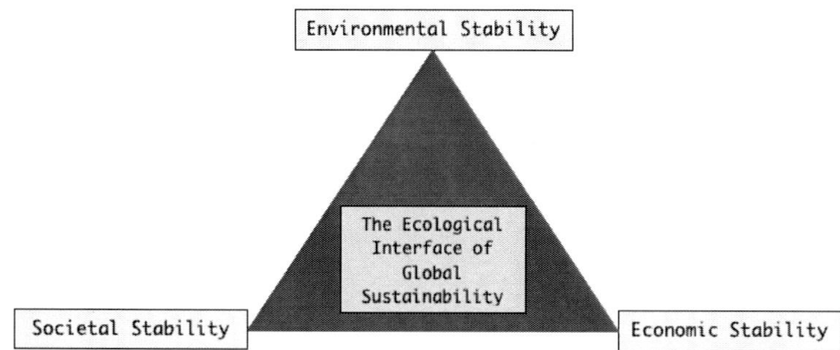

Note: The Interface of a proposed Ecological Model which includes all equal components with an aim of reaching global sustainability. Without one factor proportionately considered with the others, does not allow success in reaching sustainability.

Figure 1. The ecological model: The interface of global sustainability.

Our study

The methodology used for this narrative review includes data retrieval of the Access to Global Online Research in Agriculture (AGORA): Research4Life academic and professional peer-reviewed content database. Keyword combinations included Biodiversity conservation, (and) sustainable development, (and) emerging disease (and/or) climate-related inequities (and) disease prevention; in association with social determinants of health (or) economical determinants of health, (or) environmental determinants of health. A total of 102 papers resulted from the search. 31 papers were selected based on inclusion and exclusion qualifications. The inclusion of papers was determined by factors that were pertinent to the purpose of the review, which included a firm comparison and analysis of the direct impact of various ecological factors impacting the environmental ecosystems, which is impacting the quality of health around the world. The exclusion of papers had a publication date older than 2012 and those that were outside the scope of the study.

Discussion

There are many studies that focus on the drivers of disease, but not on the ecosystem thresholds and the ‘warning signs’ that specific environments and species provide (2). There remains a lack of understanding of the changes in climate and their effects on disease drivers (3). These gaps of instability, coupled with the warnings from the environment within the ecological interface, are reasons for intervention and support from the appropriate stakeholders (1). The economic impact of non-communicable disease includes direct and indirect costs comprising individual medical care, loss of income, the value of life, associated mental health conditions, and industry loss of the workforce (4, 6). The combination of communicable and non-communicable disease burdens puts significant pressures on individual families, communities, healthcare systems, and the economy, which is a major concern for public health (7, 11).

Proposing solutions to risk potential regarding emerging diseases is at the forefront of preserving the health of humans for generations to come (12). Many propositions to improve ecosystem sustainability have been made, such as technological and medical advances, to curb the potential impacts of disease (10). However, ecological change is an interface that requires sensitive consideration in its involvement in emerging disease occurrence (8). To increase stakeholder interest and potential buy-in, these tools can have multi-systemic surveillance that involves an effective assessment of the planet's environmental ecosystems, societal stability, and economic stability combined (9).

The largest population affected by disease are individuals living in communities of low socioeconomic status (SES), these indicators include poverty, education level, and crowded communities (1, 6, 7). The population-based surveillance of low SES measures can predict disease incidence and mortality rates to create solutions for health equity (1). There are many links to the risk of emerging disease correlated with the level of socioeconomic status (1, 13). Climate variability may interact with various drivers that ultimately alter the patterns of human migration, including political, economic, and environmental drivers (13). A lack of food and water security is associated with climate change, human migration, and health outcomes (13). Conversely, there is limited information on the direct relationship between climate change and health outcomes, but indications have been made that there is an association between environmental changes, emerging diseases, and societal determinants (3, 13). Therefore, it is beneficial to classify and limit critical tipping points of global health status by preventing risky combinations of drastic climate, societal and economic damages (3).

For accurate health policy discussions, it is imperative to receive information on the spread of disease in a reliable and timely manner (14). In regions of the world with the highest poverty rates, the greatest impact of climate change, and reduced access to food and water sources, for example sub-Saharan Africa, there is an increased percentage of death (76%) of communicable and non-communicable diseases compared to their global counterparts (15). The distribution of communicable diseases (infectious disease, zoonotic emerging/re-emerging disease) is higher in countries with less human development, while still being impacted by concurrent non-communicable diseases, furthering the impact toll on health at a global scale (7, 11). Surveillance studies show that in the last 10 years there has been a shift towards more non-communicable, chronic conditions (respiratory/pulmonary diseases, kidney diseases, cancers, diabetes, etc.) (6, 11, 14). There is a decline in global infectious diseases due to medical advances and effective socio-interventions; however, these implementations mask fundamental changes in accordance with the increase in chronic causes of death rather than acute infectious disease mortality (14). There is a correlation between increased population size, community education, health policy, consequences of unsustainable growth, and the lack of political support for evaluating fundamental ecological causes of disease (15).

Climate change is one of the greatest threats to human health and has a direct impact on health through air quality, sea-level rise, and influences on drinking water resources and food production systems (12, 16). Studies show that, if not mitigated soon, climate change will cause an increase in the length of transmission seasons, and geographical range, and cause a 'reshuffle' of animal species distribution by forced migration along with their diseases (16). For example, Amazon deforestation has an impactful role on the Earth's global temperatures and intensification of extreme weather changes (17). Deforestation is a direct driver of emerging infectious diseases and the spread of disease vectors (12, 16, 17). Urbanizing environmental changes cause fluctuation in climate, forced human and animal migration, and loss of the region's natural biodiversity (18). Thus, furthering the potentially devastating impact of emerging and re-emerging diseases on several global factors (12, 16-18).

Many challenges exist within attempts to mend social structure to promote global ecological changes. Political unrest delays environmental improvements and there are positive effects of human development and political stability on the environment (19). An ecological model was tested for the Tuojiang River Basin in China, which demonstrated feasibility and practicality and was found more reliable than previous models in balancing social stability, economic construction, and ecological function (20). This case study is an example of what is possible when stability in all aspects of an ecological interface is approached as equal participating components to an ecological concern (20). Even the global pandemic of COVID-19 has shown social inequities around the world (21). The United Nations and the World Health Organization provided that living near ecosystem-rich environments is associated with positive impacts on health outcomes; however, 55% of the world's population live in urban settings with limited contact with the natural environment (22, 23). With greater population density in low-income communities, more people are sharing less space-disproportionally impacting disease transmission, reduced access to healthcare, and less space for the positive effects of the surrounding natural environment (19, 21).

Companies around the world exert environmental pressures that drive the loss of biodiversity and further deteriorate the planet's sustainability (24). A business's ecological footprint requires consideration to assess its impact on biodiversity loss, and when companies exceed an ecological budget, the implications include ecological burden (25). It is the responsibility of government officials to act and intervene in development to achieve environmental sustainability and still gain economical advances (25). The lack of business ecological footprint assessment is negligent towards sustainability of resources in the future; and an increased risk of disease as environmental pressures drive the change of ecosystems, and the biodiversity within them, ultimately drives future health implications (26). There are ecological damage compensation frameworks in early development and estimations of ecological damage under current management regimes to develop a standard for sustainable development (24). The challenges involve the implementation of a standard calculation of the cost of ecological destruction, consistent management of the standards set, and the ability to hold businesses accountable for adequate compensation of damages (24). The lack of momentum for policy change, at a governmental or at an individual company level, to make climate-forward changes is a reluctance to modify conventional core objectives with the fear of losing financial stability (27). The technology for comprehensive monitoring of the genetic diversity of wild species, which would provide necessary information to conservation managers, exists and is affordable; however, they are not being utilized to their potential (9).

To achieve minimization of the risk of emerging diseases, first, there must be an improvement in gap indicators, screening, mitigation, sustainability objectives, and incentives for institutions to participate (10, 27-29). At a national level for the United Nations Environment Program (UNEP), only half of the nations are reporting or measuring trends of environmental changes; it is shown to not be sufficient in the representation of assessing unsustainable trends (28). Indicators, however, are the only tools that are required then to be used by decision-makers for these indicators to stimulate change. Sustainability measurement is present, although not sufficient to be a priority in national decision-making (28). The best indicators need to be complemented by other measures to strengthen their weight of importance and earn the confidence of reliability (28). An improvement in conceptualizing the link between socioeconomic development and link to ecosystem response is desperately needed due to greater pressure placed on ecosystems creating imbalances that will remain for years (29). Companies that understand how climate change links to operational frameworks and contribute to a "greener" approach to mitigate climate-related risks (29).

Enhanced environmental protection is beneficial yet can still have detrimental effects on biodiversity conservation as economic development continues to grow exponentially and unsustainably (30). Large-scale economic and environmental changes can affect multiple rich, diverse, biotic species (30). These seemingly small interruptions in local biodiversity go initially unnoticed, although they have a great impact on the stability of the remaining biotic environment (8, 30). When a significant portion of an ecosystem is altered, it mends the various species' roles in the environment, inviting invasive species, interrupting food networks, and causing ecosystem dysfunction, resulting in the potential loss of a fundamental species (8). Urbanization interferes with the proper functioning of ecological networks, but when a sustainable development approach is used there are environmental benefits that a balanced ecosystem can provide (18). A balanced ecosystem with a biodiverse-rich environment can provide carbon sequestration, particulate matter retention, reduction of surface runoff, and have a direct impact on the well-being and health of the community members (18).

The present climate crisis has the potential to impede global economies and requires urgent policy interventions (12). There is a multi-disciplinary field associated with recognizing the developmental origins of health and disease and the impact that environmental factors have on early life health problems and adulthood disease (31). Setting sustainable development goals, using existing data, and utilizing the help of stakeholders can have an impact on the improvement of overall health in many communities around the world (18, 30, 31). Assessment of environmental damages, changes, and loss of biodiversity and ecosystem stability alone does not suffice as a method of ecosystem stability (12, 30). Acknowledging that there is a required shift in decision-making to prioritize protecting, enhancing, correcting, and preventing continuous environmental destruction that has led to the damaging consequences that many communities globally have been confronted with (12, 21, 31). This shift in attention to an ecologically stable environment can equate to saving millions of lives and a reduction in economic losses (4, 5, 31).

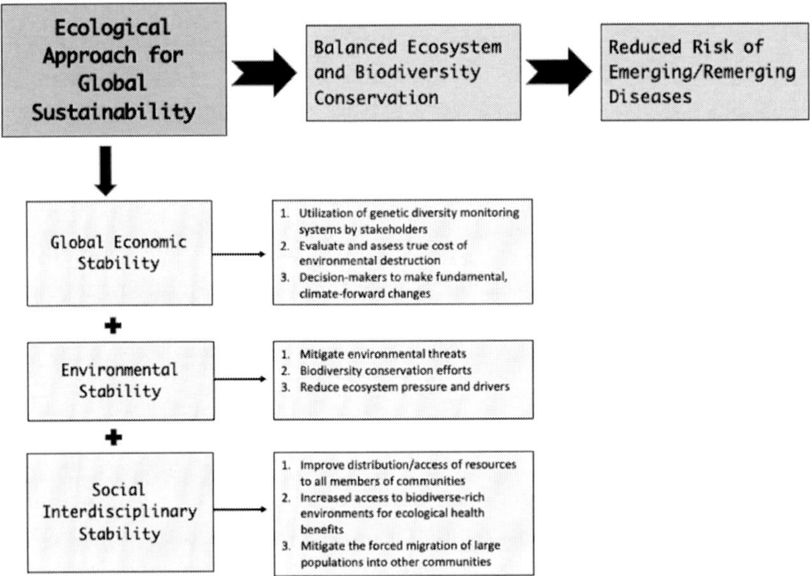

Note: The Ecological approach for global sustainability via the interface of economic, environmental, and societal stability. In working towards a more balanced ecosystem through biodiversity conservation to the attempt to reduce the risk of emerging and re-emerging disease impact on the global health of plants, animals, and people.

Figure 2. The ecological approach flow chart.

Conclusion

The proposed method of filling the research gaps of the true cost of global trends of unsustainable development includes careful consideration of environmental stability and its multiple contributing factors. Further evaluation of the direct association between loss of biodiversity, emerging disease, economic burden, health equity, and environmental quality is required for long-term benefits of environmental stability as the source for overall health improvement for plants, animals, and people alike. Stakeholders can use existing data, suggested strategies, and economical contributions to change the culture of human development into a more sustainable, conservative, and stable ecological system. The goal is to manage inevitable economic and developmental changes thoughtfully by protecting and prioritizing vulnerable ecosystems and minimizing unsustainable ecological changes as the means for emerging disease prevention and control. As successful developers of this world, there is a responsibility to maintain the planet's resources and biomes for generations to come. It is long overdue to take accountability for the damages that have occurred and to direct global growth in a sustainable direction to minimize devastating environmental effects and infectious disease occurrence, and to prevent the cause of further economic and health decline. An effective model to be used for improving environmental stability should be an equal contender among global economic and societal stability regarding the improvement of health equity and mitigation of emerging disease potential at the source.

References

(1) Hadler JL, Vugia DJ, Bennett NM, Moore MR. Emerging infections program efforts to address health equity. Emerg Infect Dis 2015;21(9):1589-94.

(2) Heffernan C. Climate change and multiple emerging infectious diseases. Vet J 2018;234:43-7.

(3) Bell N, Wilkerson R, Mayfield-Smith K, Lòpez-De Fede A. Community social determinants and health outcomes drive availability of patient-centered medical homes. Health Place 2021;67:102439.

(4) Bloom DE, Cafiero ET, Jané-Llopis E, Abrahams-Gessel S, Bloom LR, Fathima S. The global economic burden of noncommunicable diseases. Geneva: World Econ Forum, 2011.

(5) Kumar R, Narain JP, Dikid T. Noncommunicable Diseases: Health burden, economic impact and strategic priorities. In: Narain JP, Kumar R, eds. Textbook of chronic noncommunicable diseases: The health challenge of 21st century. New Delhi, India: Jaypee Brothers Medical Publishers, 2016:1-18.

(6) Rao SS, Singh RB, Takahashi T, Juneja LR, Fedacko J, Shewale AR. Economic burden of noncommunicable diseases and economic cost of functional foods for prevention. In: Singh RB, Watson RR, Takahashi T, eds. The role of functional food security in global health. New York: Academic Press, 2019:57-68.

(7) Emadi M, Delavari S, Bayati M. Global socioeconomic inequality in the burden of communicable and non-communicable diseases and injuries: An analysis on global burden of disease study 2019. BMC Public Health 2021;21(1):1771.

(8) Cagua EF, Wootton KL, Stouffer DB. Keystoneness, centrality, and the structural controllability of ecological networks. J Ecol 2019;107(4):1779-90.

(9) Pärli R, Lieberherr E, Holderegger R, Gugerli F, Widmer A, Fischer MC. Developing a monitoring program of genetic diversity: What do stakeholders say? Conserv Genet 2021;22(5):673-84.
(10) Meißner N. The incentives of private companies to invest in protected area certificates: How coalitions can improve ecosystem sustainability. Ecologic Econ 2013;95:148-58.
(11) Vandenberghe D, Albrecht J. The financial burden of non-communicable diseases in the European Union: A systematic review. Eur J Public Health 2020;30(4):833-9.
(12) Intergovernmental Panel on Climate Change. Global Warming of 1.5°C- Summary for Policymakers. IPCC 2018. URL: https://www.ipcc.ch/sr15/chapter/spm/.
(13) Schwerdtle PN, McMichael C, Mank I, Sauerborn R, Danquah I, Bowen KJ. Health and migration in the context of a changing climate: A systematic literature assessment. Environ Res Lett 2020;15(10):103006.
(14) GBD 2013 Mortality and Causes of Death Collaborators. Global, regional, and national age–sex specific all-cause and cause-specific mortality for 240 causes of death, 1990–2013: A systematic analysis for the global burden of disease study 2013. Lancet 2015;385(9963):117-71.
(15) Sigfrid L, Bannister PG, Maskell K, Regmi S, Collinson S, Blackmore C. Addressing political, economic, administrative, regulatory, logistical, ethical, and social challenges to clinical research responses to emerging epidemics and pandemics: A systematic review. Lancet 2019;394(S86).
(16) Caminade C, McIntyre KM, Jones AE. Impact of recent and future climate change on vector-borne diseases. Ann NY Acad Sci 2018;1436(1):157-73.
(17) Ellwanger JH, Kulmann-Leal B, Kaminski VL, Valverde-Villegas JM, Veiga ABGD, Spilki FR. Beyond diversity loss and climate change: Impacts of amazon deforestation on infectious diseases and public health. An Acad Bras Ciênc 2020;92(1):e20191375.
(18) Wiesel PG, Dresch E, de Santana ERR, Loboestan EA. Urban afforestation and its ecosystem balance contribution: A bibliometric review. Manag Environ Qual: Int J 2021;32(3):453-69.
(19) Mrabet Z, Alsamara M, Mimouni K, Mnasri A. Can human development and political stability improve environmental quality? New evidence from the MENA region. Econom Model 2021;94:28-44.
(20) Clapp J, Moseley WG, Burlingame B, Termine P. Viewpoint: The case for a six-dimensional food security framework. Food Policy 2022;106:102164.
(21) Geary RS, Wheeler B, Lovell R, Jepson R, Hunter R, Rodgers S. A call to action: Improving urban green spaces to reduce health inequalities exacerbated by COVID-19. Prev Med 2021;145:106425.
(22) United Nations. World urbanization prospects: The 2018 revision. New York: UN, 2019. URL: https://www.un.org/development/desa/pd/content/world-urbanization-prospects-2018-revision.
(23) World Health Organization. Urban green space interventions and health: A review of impacts and effectiveness. Geneva: WHO Reg Office Eur, 2017. URL:

https://www.who.int/europe/publications/m/item/urban-green-space-interventions-and-health--a-review-of-impacts-and-effectiveness.-full-report.

(24) Rao H, Lin C, Kong H, Jin D, Peng B. Ecological damage compensation for coastal sea area uses. Ecologic Indicators 2014;38:149-58.

(25) Wolff A, Gondran N, Brodhag C. Detecting unsustainable pressures exerted on biodiversity by a company: Application to the food portfolio of a retailer. J Clean Product 2017;166:784-97.

(26) Di Minin E, Soutullo A, Bartesaghi L, Rios M, Szephegyi MN, Moilanen A. Integrating biodiversity, ecosystem services and socio-economic data to identify priority areas and landowners for conservation actions at the national scale. Biologic Conserv 2017;206:56-64.

(27) Dikau S, Volz U. Central bank mandates, sustainability objectives and the promotion of green finance. Ecologic Econom 2021;184:107022.

(28) Dahl AL. Achievements and gaps in indicators for sustainability. Ecologic Indicator 2012;17:14-9.

(29) Lu Y, Liu M, Zeng S, Wang C. Screening and mitigating major threats of regional development to water ecosystems using ecosystem services as endpoints. J Environ Manage 2021;293:112787.

(30) Pagani-Núñez E, Xu Y, Yan M, He J, Jiang Z, Jiang H. Trade-offs between economic development and biodiversity conservation on a tropical island. Conserv Biol 2022 Mar 25. doi: 10.1111/cobi.13912.

(31) Norris SA, Daar A, Balasubramanian D, Byass P, Kimani-Murage E, Macnab A. Understanding and acting on the developmental origins of health and disease in Africa would improve health across generations. Glob Health Action 2017;10(1):1334985.

Chapter 12

Anisakiasis and anisakis larvae in Japan: The prevalence and risk of anisakis infections

Arashi Nakashima, DVM
Sara Elzibak, BSc
and Satesh Bidaisee*, DVM, MSPH, EdD
Public Health and Preventive Medicine, St George's University, Grenada, West Indies

Abstract

Anisakiasis is a parasitic disease caused by anisakis penetrating the stomach or intestinal wall. Anisakis is a parasite designated as a causative agent of food poisoning by the revision of the Food Sanitation Law in 1999. Furthermore, in 2013, items related to parasites such as anisakis and kudoa were added to the list of etiological agents and types in the food poisoning incident form; and anisakis food poisoning came to be counted separately in the food poisoning statistics. As a result, the incidence of anisakis food poisoning and the causative foods have become more apparent, and the number of reported cases of anisakis food poisoning has gradually increased- reaching 469 cases in 2018, more than double the number of the previous year. In this article, we review the articles published in Japan and present the situation of anisakis food poisoning in Japan, and the parasitic status of anisakis in fish and shellfish and discuss specific methods of preventing and coping with infection.

* ***Correspondence:*** Satesh Bidaisee, Professor, Public Health and Preventive Medicine, St George's University, St George, Grenada, West Indies. Email: sbidaisee@sgu.edu

In: Public Health: Intersection of Health, Humans, Animals and the Environment
Editors: Satesh Bidaisee and Joav Merrick
ISBN: 979-8-89530-443-3

Introduction

Parasites are one of the most important diseases in medicine, veterinary medicine, and public health. Zoonoses have long been a challenge and public health, which is a primary prevention, has been considered more effective than medicine, a secondary prevention. However, it is very difficult to deal with the many parasites that exist through public health prevention alone. As a result, we can see regular outbreaks of parasitic diseases around the world. At times, there have been cases that developed into epidemics. Anisakiasis is one such case where many infections have been confirmed, especially in areas where a lot of raw fish is consumed. It is quite possible to prevent social and economic losses by understanding the importance of public health. If parasitic diseases are to be prevented from a public health perspective, education of people, including correct cooking methods and checking of details, is considered most effective.

The sensitization rate of anisakis simplex spp. is increasing worldwide and is having a significant impact on health care systems (1). Anisakis disease has increased with the development of the distribution system; it is currently estimated that there are more than 500 cases per year since the first report in 1965. The number of cases is considered the highest, and anisakiasis in Japan is designated as a causative agent of food poisoning among foodborne parasitic diseases under the Food Sanitation Law. Mackerel is the most common food-causing anisakiasis in Japan (2). In 2017, it was estimated that 42.0% of anisakis food poisoning was caused by mackerel. Raw edible saury, squid, and bonito have also been reported (3). Therefore, in this report, we describe the discovery of anisakiasis caused by anisakis and its history up to the present, outline the morphological and molecular biological identification techniques of anisakis, and discuss the parasitic status of anisakis in marine products and its risk of infection in humans.

Our review

A systematic review of peer-reviewed Japanese-language literature for anisakis spp. sensitization prevalence data was conducted through a search of CiNii databases (Japanese scientific paper database). Initially, free text words representing the broad concept of "anisakis" were used to identify the keywords for subject searching. Then, a combination of MSH terms and free text words were arranged in the following research string with "or" and "and" logical operators: Anisakis "and" (prevalence "or" epi- demiology) "and" (allergy "or" hypersensitivity "or" immunization "or" sensitivity "or" sensitization "or" ELISA "or" skin prick test "or" ImmunoCAP "or" Immunoblot "or" diagnostic techniques). Reference lists of the articles included in the analysis and other papers related to the issue were hand-searched for further, possibly relevant papers until no fresh material was discovered (1).

All articles that met the following criteria were screened and then evaluated for eligibility: peer-reviewed manuscripts published between 2010 and December 2020, reporting prevalence estimates of anisakis sensitization and describing the population of interest, the technique used, and the number of people tested. Cross-sectional, prospective, or retrospective studies were eligible to be reported, and there were no restrictions on age or population type. In addition, official reports from national institutes were included since there were not enough accessible articles present in the database (1).

Data was extracted using the Data Extraction Microsoft Excel sheet. Criteria employed for data extraction included author, year of publication, year of study, study location, study design, statistical measures, study setting, sample size, characteristics of study participants as age and female/male ratio, and diagnostic techniques and to define allergy and relative prevalence estimates with anisakis sensitization.

Discussion

The characteristics of the seven studies examined in the qualitative analysis are summarized in Table 1. One study followed systematic review, one study followed the cross-sectional design, and the remaining studies were designed as a cohort study. Two studies defined specific locations (4, 5). Except for one study, six studies were considered medium-high quality since the study population was insufficient (5).

The larvae parasitizing a wide variety of marine fish, such as cod, herring, horse mackerel, bonito, sardine, saury, salmon, trout, and squid, are the source of human infection (see Table 2). Although textbooks usually mention that tuna is not a parasite, it was unexpected to learn that the parasite is found in bigeye tuna (relatively small bluefin tuna weighing less than 20 kg). As far as could be determined, the species' with no problem at all was squid, rockfish, flounder, and flatfish. This distribution is considered as "not very mobile"; therefore, their risk of infection is potentially zero (3).

Table 1. Characteristics of seven studies

Author	Year	Type	Period	Study population	Mean age	Location
Jun. M	2012	Systematic review	not provided	not provided	not provided	Japan
Food Safety Commission of Japan	2018	Cohort study	2008-2016	not provided	not provided	Japan
Toshio. A (4)	2014	Cohort study	2011-2012	44	54.5	Middle East Japan
Takashi. F	2020	Cross sectional	2018	25	not provided	Japan
Takashi. B	2020	Cohort study	2017-2020	128	Japanese sardines were target	Japan
Jun. S (7)	2020	Cohort study	2011-2018	1254	not provided	Japan
Mizuho. S (5)	2011	Cohort study	2009-2011	2	not provided	Tochigi Pref. Japan

Table 2. The possibility of contracting anisakis from eating raw marine fish and shellfish

Possibility of harboring infected larvae	Species	Distribution
High	Cod, herring, cherry salmon, red sea bream, bonito	Mostly distributed or migratory in northern Japan.
Moderate	Horse mackerel, mackerel, squid	Coast of Japan, with seasonal migration
Low	Pacific saury, sardine, bigeye tuna	Coast of Japan, with seasonal migration
Very low	Tuna	Warm waters, great returnability
Nothing	Squid, rockfish, flounder, flatfish	Not very mobile

The reason behind the sharp increase in statistics is thought to be the spread among medical professionals of the recognition that anisakis infection should be reported as food poisoning, and not a sharp increase in recent years. Another factor thought to have contributed to the increase in the number of cases is that the development of low-temperature distribution systems for fresh food has made it easier to eat fresh raw fish and shellfish landed in remote areas. According to statistical data, mackerel is the most commonly known causative food, with saury, skipjack tuna, and horse mackerel being the others, and food poisoning outbreaks tended to occur mostly after summer (6).

However, in 2018, perhaps due to the bountiful catch of skipjack tuna, there was a sharp increase in the number of incidents involving skipjack tuna as the causative food, with most cases occurring in spring and summer (see Table 3).

Table 3. Number of incidents of anisakis food poisoning

Year	Number of positives
2010	28
2011	34
2012	65
2013	88
2014	79
2015	127
2016	124
2017	230
2018	468
2019	328
Total	1,571

Mackerel is the most common causative food of anisakiasis (including cases without food poisoning) in Japan, especially in Tokyo, where homemade "shime mackerel" is often produced in restaurants, supermarkets, and retail stores eaten as a cause. 38.6% of the cases in Tokyo in 2011 and 2012, and 42.0% in 2017, were estimated to be caused by "shime mackerel." Due to the high parasite rate of anisakis in mackerel, 70% of restaurants and seafood retailers that serve homemade shime mackerel use refrigerated mackerel. In addition, the trend from 2013 to 2017 is thought to be due to the increase in the number of anisakis food poisoning cases caused by saury in september and october. However, in 2018, many anisakiasis cases caused by the first bonito were reported in april and may nationwide (7).

In the cases of anisakis in Tokyo between 2011 and 2018, 94% of them were identified as anisakiasis simplex sensu stricto and 4.4% as anisakiasis pegreffii. This well reflects the report that the transfer rate of anisakiasis simplex into the muscle of mackerel is more than 100 times higher than that of anisakiasis pegreffii, suggesting that anisakiasis simplex is likely to be the causative parasite of anisakis food poisoning as a result of the high transfer rate of anisakiasis simplex into the muscle of fish. In addition, in 2018, the number of anisakiasis simplex in Tokyo, 23 requests were received for testing of cases of anisakiasis presumed to be caused by eating bonito in Tokyo, and all 31 specimens tested were identified as anisakiasis simplex. Several cases of pseudoterranosis caused by pseudoterranova larval nematodes have been reported in Tokyo, where seals and Steller's sea lions are the final hosts, and fish and shellfish are the waiting hosts. Pseudoterranova disease is often reported as anisakis food poisoning under the Food Sanitation Law. However, unlike anisakiasis, this disease is characterized by mild abdominal pain and intense vomiting, and nausea. Pseudotelanosis is thought to be rare in Japan, except in Hokkaido (7).

According to the research from Suzuki, which investigated the cause of the increase in anisakis food poisoning caused by bonito in 2018, a survey was conducted for anisakis parasites in bonito caught at six sites on the Pacific Ocean side (off Chiba to Miyagi prefectures) in August to November 2018; it was conducted by separating the viscera and muscle parts immediately after catching and landing, in order to elucidate whether anisakis migrates into the muscle before or after catching (7).

As a result of the survey, a total of 273 anisakis larvae (251 from the internal organs of 30 bonito and 22 from the ventral muscles of 2 bonito) were detected in 30 bonito, gutted immediately after catching, and 197 of them were identified as anisakiasis simplex (see Table 4). On the other hand, in the 30 fish gutted immediately after landing, 212 anisakis third instar larvae were detected in the 28 guts and 13 in the ventral muscles of 8 fish, for a total of 225 individuals, 199 of which were identified as anisakiasis simplex. As a result, it became clear that anisakis had already migrated into the muscle before the fish were caught (see Table 4). The relative number of anisakiasis simplex parasites per fish (total number of anisakiasis simplex detected/number of skipjack tuna examined) was 6.6 (197 fish/30 fish), and the same number was found after landing (199 fish/30 fish). The relative parasitism of anisakiasis simplex was 1.5 individuals (39 individuals/26 fishes), which was less than 1/4 of the relative parasitism in august-october 2018, according to the results of surveys conducted by the center in 2012-2016. The relative number of anisakiasis simplex parasites was 1.5 (39 individuals/26 fishes), less than 1/4 of the relative number in august-october 2018.

Interviews on the migration routes of skipjack tuna, the fishing areas at the landing sites, and the distribution status of skipjack tuna after landing were conducted. In 2018, a large number of skinny skipjack tuna (around two kg), usually caught in april in the Ogasawara Islands and the Izu Islands, were caught in the same waters of the Izu Islands fed heavily on krill with more fat than usual. It may be because some of the caught bonito preyed on krill and anchovy, which are intermediate and standby hosts of anisakis, for an extended period in the seas around Japan, as the seawater temperature remained higher than usual due to the large meandering of the Kuroshio Current after september 2017. Therefore, under higher-than-usual seawater temperatures, skipjack tuna that had sufficiently fed on krill and anchovy were caught in the same sea area and distributed nationwide, resulting in food poisoning by anisakis occurring at the same time in Tokyo, Fukushima, and Miyazaki prefectures (7).

Table 4. Detection of anisakis in Bonito during different periods of visceral removal in 2018

Time to remove guts	Test number	Positives	Anisakis simplex	Anisakis physeteris	Number of anisakis		
					GI	Ventral muscle	Dorsal muscle
Immediately after the catch	30	30	197	57	251	22	0
after the land	30	28	199	15	212	13	0
total	60	58	396	72	463	35	0

Conclusion

The parasite itself is found in marine fish and marine mammals worldwide and is a parasitic nematode that lives in the marine world. In the past, infections were often observed in the Netherlands among fishers who ate dancing herring; however, after this practice was banned and only refrigerated or frozen-thawed herring was permitted to be eaten, the number of cases decreased dramatically (8).

Because of its customs of enjoying various kinds of fish sashimi, Japan is still the country with the highest number of infected people in the world. It is said that people do not come to the Department of Parasitology in medical schools for consultation because it is so common, and the number of cases is so large. In this paper, the critical clinical presentations were the focus. The possibility of the spread of infection in Japanese restaurants in Europe and the United States should be considered when formulating countermeasures. In

Europe, there are many opportunities to eat sushi and yakitori on the street, even in large cities. While it is excellent that traditional Japanese food such as sushi is spreading internationally, there is a concern that the possibility of parasitic infections such as anisakis may be transmitted at the same time. In Japan, it is not well known that anisakis can be detected in mackerel sushi, but it is necessary to urge caution overseas as well (3).

The primary prevention is to avoid eating raw fish, which is the source of infection. It is essential to avoid eating raw marine fish as much as possible. However, it is almost impossible to do so in Japan. Fish that has been frozen for at least two days and then thawed is safe. In Japan, it is expected that cooks will be careful when preparing the fish, but this may not be the case overseas. Avoid eating raw fish and shellfish, which are sources of infection for humans. It may be possible for foreigners who eat more meat and less fish. However, sashimi and sushi lovers are now spreading around the world. As long as fish is grilled or thoroughly heat-treated as much as possible, there is no problem of anisakis infection.

Eating raw fish that has been frozen and thawed is feasible. In research from Baba, it has been done successfully with herring in the Netherlands (6). Storage at -20°C for 24 hours or freezing for at least two full days is recommended.

Secondary prevention, such as early detection and treatment, involves immediately recalling the raw consumption of infected fish and seeking appropriate medical attention. The following symptoms, diagnosis, and treatment are the key factors to look for.

If the parasite is lodged in the gastric wall, orbital pain is observed several hours after eating. In the past, the stomach was sometimes removed, but this is not necessary at all because the larvae are not likely to survive for more than a week. The environment of a human stomach is considerably different to that of a whale, where these larvae can survive for longer. This may be due to the fact that the digestive juices in the stomach of a whale are mixed with seawater, although this should be further analyzed to understand the mechanism of survival and provide an opportunity to combat such infections. If the parasite is lodged in the intestines, severe cases may include allergic symptoms, such as an acute abdomen.

Since only larvae parasitize humans, there is no way to produce worm eggs, and of course, stool tests are meaningless. In the case of adult worms infecting marine mammals, such as dolphins, worm eggs may be found in stool samples. In other words, the presence or absence of adult worm infection can be determined, but this is the domain of veterinary medicine. Usually, anisakis larvae in the stomach are detected by endoscopy and removed with forceps. X-ray gastrointestinal fluoroscopy can also be used to detect the worm body.

A cure has yet to be developed. Anisakis parasites in the stomach can be mainly removed, but not in all cases. In some cases, it has been challenging to remove the parasites due to an abundance of parasites (4). For anisakis parasites in the small intestine, anthelmintics are necessary. Since there are some candidates, it is worthwhile to conduct a large-scale in vitro study first. The larvae of anisakis can survive in the human gastrointestinal tract for only about one week at most, and some believe up to one month. In other words, the worm's body is killed and absorbed, so it is possible to avoid using anthelmintics. When infected, the patient suffers pain to varying degrees. It is often like an allergic reaction to repeated infection (9).

It is expected that patients suffering from severe stomach pains, diarrhea, and abdominal pains after eating raw marine fish may be considered for possible infection with this worm in the future. It is vital to educate pharmacists to give appropriate advice to patients who come to purchase stomach painkillers. Due to changes in social conditions, there still seems to be a misconception that parasites are no longer present in Japan. There may not be a general awareness that Anisakis is a parasite. In addition, since "parasitology" is exceptionally inadequate in microbiology education in the Faculty of Pharmacy, there are cases where knowledge of anisakis cannot be acquired. Pharmacists, who are defined as medical professionals, must be aware of this fact.

In some cases, appropriate remedial courses may be necessary. The pharmacist's role as a member of the health care team should be considered in more detail. As a member of the health care team, the pharmacist will need to be aware of the possibility of infection in patients who are suffering from severe diarrhea and abdominal pain after eating raw marine fish throughout the country and abroad. In pharmacy, the creation of anthelmintic drugs for Anisakis, which has not yet been developed, is an urgent issue (7).

To reiterate, due to changes in social conditions, some people still have the misconception that parasites have disappeared in Japan. However, it is essential to correct this misconception through educational activities and reduce economic and social losses. The primary prevention of anisakis infection is possible with accurate knowledge. Such prevention can significantly reduce the current social and economic losses in Japan. This knowledge is also helpful for secondary prevention.

References

(1) Mazzucco W, Raia DD, Marotta C, Costa A, Ferrantelli V, Vitale F, et al. Anisakis sensitization in different population groups and public health impact: A systematic review. PLoS One 2018;13(9):e0203671.
(2) National Institute of Infectious Diseases. Foodborne helminthiases in Japan. Infect Agents Surveil Rep 2017;38(4):69-70. URL: https://www.niid.go.jp/niid/en/iasr-vol38-e/865-iasr/7225-446te.html.
(3) Suzuki J. Food poisoning caused by juvenile nematodes and Anisakis. J Vet Epidemiol 2019;23(1):66-7. URL: https://doi.org/10.2743/jve.23.66.
(4) Arai T, Akao N, Seki T, Kumagai T, Ishikawa H, Ohta N, et al. Molecular genotyping of anisakis larvae in middle eastern Japan and endoscopic evidence for preferential penetration of normal over atrophic mucosa. PLoS One 2014;9(2):e89188.
(5) Shimada M, Komatsumoto S, Kirinoki M, Chigusa Y, Matsuoka H. Characteristics of patients with bee sting, centipede bite, or viper bite treated at Ashikaga red cross hospital in Tochigi Prefecture, Japan between 2009 and 2011. Med Entomol Zool 2012;63(2):103-7.
(6) Baba T. Molecular identification and prevalence of Anisakis larvae in Japanese sardine Sardinops melanostictus from Japanese waters. Japan J Fisheries Sci 2021;87(1):52-4.
(7) Suzuki J. Food poisoning caused by anisakis larvae and its causative foods in Japan. Jap J Food Microbiol 2020;37(3):122-5.
(8) Sugiyama H, Morishima Y, Kagawa C, Araki J, Iwaki T, Ikuno H, et al. Current incidence and contamination sources of ascariasis in Japan. Shokuhin Eiseigaku Zasshi 2020;61(4):103-8.
(9) Murata R, Suzuki J, Sadamasu K, Kai A. Morphological and molecular characterization of Anisakis larvae (Nematoda: Anisakidae) in Beryx splendens from Japanese waters. Parasitol Int 2011;60(2):193-8.
(10) Kamada Y. Anisakis, anisakis food poisoning, anisakis allergens. Food Analysis Technol Center. URL: http://www.mac.or.jp/mail/210201/02.shtml.
(11) Bao M, Pierce GJ, Pascual S, González-Muñoz M, Mattiucci S, Mladineo I, et al. Assessing the risk of an emerging zoonosis of worldwide concern: Anisakiasis. Sci Rep 2017;7(1):43699.
(12) Deardorff TL, Kent ML. Prevalence of larval anisakis simplex in pen-reared and wild-caught salmon (salmonidae) from Puget Sound, Washington. J Wildl Dis 1989;25(3):416-9.
(13) Gomes TL, Quiazon KMA, Kotake M, Itoh N, Yoshinaga T. Anisakis spp. in fishery products from Japanese waters: Updated insights on host prevalence and human infection risk factors. Parasitol Int 2020;78:102137.
(14) Kong Q, Fan L, Zhang J, Akao N, Dong K, Lou D, et al. Molecular identification of Anisakis and Hysterothylacium larvae in marine fishes from the East China Sea and the Pacific coast of central Japan. Int J Food Microbiol 2015;199:1-7.
(15) Nieuwenhuizen NE. Anisakis - immunology of a foodborne parasitosis. Parasite Immunol 2016;38(9):548-57.

(16) Nogami Y, Fujii-Nishimura Y, Banno K, Suzuki A, Susumu N, Hibi T, et al. Anisakiasis mimics cancer recurrence: two cases of extragastrointestinal anisakiasis suspected to be recurrence of gynecological cancer on PET-CT and molecular biological investigation. BMC Med Imaging 2016;16(1):31.
(17) Quiazon KMA, Yoshinaga T, Ogawa K. Distribution of Anisakis species larvae from fishes of the Japanese waters. Parasitol Int 2011;60(2):223-6.
(18) Quiazon KMA, Zenke K, Yoshinaga T. Molecular characterization and comparison of four anisakis allergens between anisakis simplex sensu stricto and anisakis pegreffii from Japan. Mol Biochem Parasitol 2013;190(1):23-6.
(19) Suzuki J, Murata R, Hosaka M, Araki J. Risk factors for human Anisakis infection and association between the geographic origins of scomber japonicus and anisakid nematodes. Int J Food Microbiol 2010;137(1):88-93.
(20) Wiwanitkit S, Wiwanitkit V. Anisakiasis in Southeast Asia: A story of new tropical disease? Asian Pacific J Trop Biomed 2016;6(5):382-3.
(21) Zolfaghari Emameh R, Purmonen S, Sukura A, Parkkila S. Surveillance and diagnosis of zoonotic foodborne parasites. Food Sci Nutr 2017;6(1):3-17.

Chapter 13

Survey related to Leatherback sea turtle consumption of people 18 years or older in Grenada

Caitlin Birky, DVM, MPH
Sara Elzibak, BSc
and Satesh Bidaisee*, DVM, MSPH, EdD
Public Health and Preventive Medicine, St George's University, Grenada, West Indies

Abstract

The Leatherback sea turtle (*Dermochelys coriacea)* is known for traveling large distances and choosing long stretches of beach in warmer climates to nest and lay their eggs, one of which is in Grenada. These animals are on the Endangered Species list in the United States due to multiple factors, such as poaching and habitat destruction, that place them at risk. The consumption of Leatherback sea turtle eggs and meat are hazardous to both turtles, as it further emphasizes their endangerment, and the people who consume them, as they may contain bacteria that are harmful. This study is a qualitative study that used both scientific measures, used for statistical data regarding the turtles, and human interviews of the people of Grenada over the age of 18 years. It is hypothesized that most people interviewed will have participated in the practice of consuming turtle eggs and meat at some point in their life, and that they are unaware of the species position on the endangered species list. The objectives of the study are to access the level of understanding

* ***Correspondence:*** Satesh Bidaisee, Professor, Public Health and Preventive Medicine, St George's University, Grenada, West Indies. Email: sbidaisee@sgu.edu

In: Public Health: Intersection of Health, Humans, Animals and the Environment
Editors: Satesh Bidaisee and Joav Merrick
ISBN: 979-8-89530-443-3

the risks associated with this practice (consumption of turtle eggs and meat), gather a baseline of knowledge on how Grenadian's view this practice (Is it accepted, or is it frowned upon?), determine how many people participate in this practice, and find out what pathogens are found in or around sea turtle eggs to better understand how they could be a threat to public health.

Introduction

The Leatherback sea turtle (*Dermochelys coriacea)* is known for traveling large distances and is the only remaining species of the Dermochelys genus (1). They have been known to travel over 10,000 miles in one year in order to stay in warmer climates and to look for food (2). They choose long stretches of beaches in warmer climates to nest and lay their eggs, one of which is located in Grenada. While the turtles are laying their eggs, they can be at risk from predators, but also from anthropogenic activity. They are at risk of being taken advantage of, having their habitat destroyed, having artificial light interrupt their laying process, and from the contamination of plastic and other trash (3). The reasons listed are major factors of influence as to why these animals are on the International Union for Conservation of Nation's Red List as venerable and the Endangered Species list in the United States (4). Due to this status, more conservation efforts need to be implemented all year long, and not only during the nesting season, because it is known that after the season is complete, some remain close by and are at risk of being captured by fishermen (5).

Leatherback sea turtles are an important factor to include in more conservation efforts, because they are directly beneficial to Grenada. As of 2013, Grenada attributed 18% of the gross domestic product (GDP) to Industry, which accounts for the country's tourism and is an important source of income for the country (6). In the nesting season, many tourists and students at St George's University (SGU) like to participate in trips to Levera Beach to witness the female turtles come out of the ocean and find the perfect spot to lay their eggs. There are even student groups who volunteer to go and collect research and generate statistics based on the health of the laying turtle, the number of eggs laid, and then from the bacteria present on the hatched eggshells. The collection of this information is helping create a baseline of information on how these turtles play a role in public health.

Not much information has been gathered on the health risks of ingesting or handling Leatherback sea turtle meat or eggs in Grenada; two studies from Costa Rica found that *salmonella* was a major potential risk for the turtles and for humans, and that a large role of hatchling success is based on the microbial and fungal varieties found in the sand inside the nesting hole (1, 7). In Mexico, a survey was conducted to determine the public health safety and knowledge associated with the consumption of sea turtles. Their results showed that even physicians were not fully aware of the exact possible side effects that could occur from turtle consumption, and that many people may stop consuming it if they were told it was unhealthy (8).

In one study done in Grenada, the eggs were immediately swabbed for bacteria after they were lain and sand samples were taken to see what bacteria was present. The results from this experiment found *pseudomonas* spp. in abundance and many other pathogens (9). The article then addressed how this is a public health issue due to the pathogens being "associated in humans with urinary tract infections, respiratory and gastrointestinal disease, wound infections, sepsis, and meningitis." Another study points out that there is an added risk of getting heavy metal contamination from eating sea turtle meat. Some of the levels are higher than international food safety standards, and "could result in … neurotoxicity, kidney disease, liver cancer, and developmental effects in fetuses and children" (10). It was also found in a similar study that these effects can be seen throughout families due to an increase in sharing food (11).

People in Grenada are at risk of the mentioned previously due to the high prevalence of the local customs of ingesting the turtle meat and eggs, and even handling the eggs. The consequences of this practice could be severe diarrhea, severe dehydration that leads to hospitalization, and possibly toxic metal poisoning. Also, if common hand washing techniques are ignored in between handling the eggs and eating, there is a chance that bacteria will remain on the hands, and it could cause that person to become infected by different bacteria. As seen by the Grenadian study, there is already an accumulation of some antimicrobial resistant bacteria found on the outer layer of the eggs and in the surrounding sand (9). These risks are literally right outside for some people living in coastal communities around Grenada.

Hypothesis (H_1): That a majority of people interviewed will have participated in this practice at some point in their life, and that they are unaware that the Leatherback sea turtle is on the endangered species list.

Null Hypothesis (H_0): No one interviewed would have participated in consuming turtle meat or eggs.

Thesis: The consumption of Leatherback sea turtle eggs and meat are hazardous to both the turtles' place on the endangered species list and the people who consume it.

Goal: The main goal of this research project would be to lead to the cessation of consumption of Leatherback sea turtle meat and eggs.

Objectives:

1) Access the level of understanding the risks associated with this practice (consumption of turtle eggs and meat).
2) Gather a baseline of knowledge on how Grenadian's view this practice.
    a) Is it accepted, or is it frowned upon?
3) To determine how many people, participate in this practice.
4) View what pathogens are found in or around sea turtle eggs to better understand how they could be a threat to public health.

Our study

The study design for this research is a qualitative study. This was determined based on the study using an Interview as its primary source of data. For the entire population of Grenada, 100,000, the sample size also would seem small; but that is allowed when doing qualitative research. All the questions featured in the survey will help build a final interpretation of the data.

For the parameters of the study, the information was collected from the months of April 2017 to December 2019. Information was also collected by a third-party group, Ocean Spirits, which is associated with St George's University. They would send a group of students up to the beaches where the turtles are laying their nests and collect data on the sand temperatures, surrounding bacteria, number of eggs per turtle, and information on the mother. The survey retrieved the remainder of the information needed for the study by having a small number of students (4) go out to set locations around the island in order to find people who want to join the study.

The study was given to randomly selected willing participants from local fishing communities around Grenada, and the St George's bus terminal. These locations were chosen based on the proximity to the beaches where the turtles come up and make their nests, proximity to fishing communities, and the bus terminal due to the fact that people from all over the island use that public transport station. The survey is included at the end of this proposal. Before starting this survey with a participant, the researcher obtained informed consent, and that they understand their rights during and after completing the questionnaire. At any point, while answering the questions, the participant could have chosen to stop and no longer partake in the study. The form that was used as informed consent is also provided at the end of this proposal. The people recruited had to be older than 18 years old and there was no discrimination between men and women. Minors were excluded so that additional consent was not needed, and so that any bias about incomplete comprehension of the questions asked or inaccurate recall of experiences could be avoided.

The number of people needed to participate in this study was n=50. This number was calculated using this website, http://sampsize.sourceforge.net/iface/index.html#prev, as a power analyzer. The assumptions associated with the analyzer were 5.15% precision, 50% prevalence with an infinite population size, and a 95% confidence interval.

In this study the independent variables were the ones that the researchers could manipulate, so in this case they would be the participants age or location where they grew up/are currently living. The dependent variable is that the same survey will be given to every person, and the questions will not be reworded for different people. It is possible that the socio-economic status of some interviewed could have become an extraneous variable, due to ingesting turtle eggs being a free meal for some, making them more likely to participate in the practice. If the socio-economic status is taken into account, then it is less likely that it will progress into a confounding variable, and then it will not lower the validity of the study.

For keeping track of all the surveys, the use of an Excel worksheet kept track of all of the answers. The researchers will also be asking questions and inputting the data so there is less chance of misunderstanding how to complete the survey.

The current plan for the survey was to only ask the questions one time around, but if it will enhance the reliability of the study, there could always be a follow up where the same questions are re-asked to ensure that the person was remembering correctly. The method would be called the Test-retest reliability. For the survey used, Face Validity will be the main type focused on. If the purpose of the survey is clear and concise, then there will be a higher face validity value.

This study has been granted IRB approval.

The data will be analyzed using the risk ratio as a measure of association between ingesting the sea turtle meat and/or eggs and becoming ill.

Findings

In total 51 people were interviewed, 39 men and 12 women. 90% (46/51) of the surveyed individuals have lived in Grenada their entire life, and predominantly most lived in St George's parish (23/51) and St John's (16/51) parish. 76% (39/51) knew about the Leatherback sea turtles and 20/51 believed that the turtles played a role in both food and the environment. 37% (19/51) told the interviewer that they had ingested turtle meat in the past, and 100% of them ate it fully cooked, only 13% (7/51) ingested sea turtle eggs and 85% of the time they were also fully cooked. No one reported getting sick after eating any turtle meat or eggs. The data showed that it was more common to have men partake in the consumption of either the eggs or meat, and there was also a spike correlating the older age groups and the amount of people who had participated, in both men and women. The main reason contributing to who ingested it was either because they liked it or because it was a part of their culture growing up. No one disclosed needing to eat turtle meat or eggs because they had absolutely nothing else available. 19% of people knew about food borne illnesses related to Salmonella specifically and 45% knew that there were specific laws in Grenada pertaining to the time when sea turtles can be caught legally, and when it becomes illegal again. One of the last questions asked in the survey pertained to if the interviewee believed that the sea turtles helped the islands in any way, and 58% (30/51) agreed that they did have added benefit to Grenada.

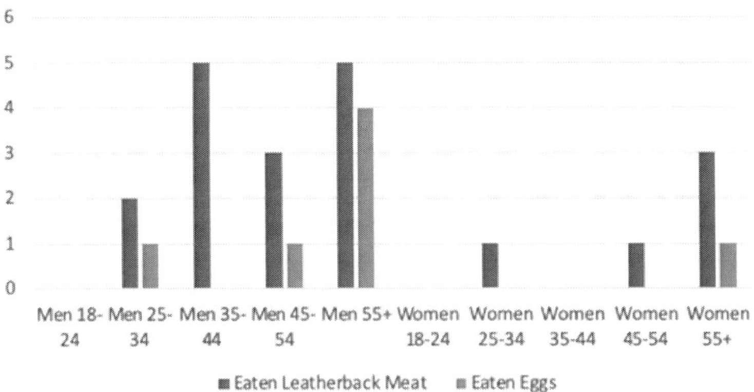

Figure 1. Number of men and women who have eaten turtle meat and eggs.

Discussion

Based on the information gained from doing this survey, it was seen that men would have eaten Leatherback sea turtles (meat/eggs) more commonly than women. In a few of the male interviews they mentioned that it was sometimes a way to prove how manly a person was. Not enough people brought this up to make it clinically significant, but that could also be linked in with their tradition. This may be an area to investigate further if the survey is ever repeated. It was more common to ingest turtle meat compared to the eggs and all of those who participated in this custom did not report any signs of being sick afterwards. There is still a possibility that individuals who handled the eggs/meat could still become carriers of bacteria like *salmonella* species, which is highly likely in reptiles (12). When a person becomes a carrier, they can shed the bacteria at different times, sometimes for months; and if they do not have consistent hygiene practices, then they have the potential to contaminate others food and spread the illness. Carriers are the most difficult area to try and prevent because these people are asymptomatic and not currently experiencing any problems. Due to the small sample size, repeating this survey in the future could potentially bring more information to light and could help make the Grenadian public aware of the risks associated with the consumption of turtle products and could also lead to better hand washing techniques and sanitary protocols. Along with the survey, if participants were also given a handout on how to decrease the risk of getting ill as well as the benefits that the Leatherback's bring to Grenada, then there would also be public education. Since one of the main reasons linked with consumption was due to cultural reasons, it might take more widespread educational talks that focus on the children and young adults in the areas on why the turtles are important to Grenada and how they should instead be protected.

Conclusion

Overall, this study accomplished the goal of finding out how common this practice was amongst the Grenadian population and what the reasoning was behind it. For most people it was a part of their heritage, while others just considered it a food they enjoyed eating. This paper does not pass any judgement, as the Grenadian people are free to participate in their own cultures and customs, but hopefully with added awareness on this issue of the

Leatherbacks' being endangered they can focus on cultivating a lasting ecological difference in this species numbers.

References

(1) Santoro M, Hernandéz G, Caballero M, García F. Potential bacterial pathogens carried by nesting leatherback turtles (Dermochelys coriacea) in Costa Rica. Chelon Conserv Biol 2008;7(1):104-8.

(2) See turtles. Sea turtle migration. URL: http://www.seeturtles.org/sea-turtle-migration/.

(3) Mathenge SM, Mwasi BN, Mwasi SM. Effects of anthropogenic activities on nesting beaches along the Mombasa-Kilifi Shoreline, Kenya. Marine Turtle Newsletter 2012;(135):14-8.

(4) See Turtles. Leatherback sea turtles. URL: http://www.seeturtles.org/leatherback-turtles/.

(5) Georges J, Billes A, Ferraroli S, Fossette S, Fretey J, Grémillet D, et al. Meta-analysis of movements in Atlantic leatherback turtles during the nesting season: Conservation implications. Marine Ecol Prog Series 2007;338:225-32.

(6) The Economy. Gov Grenada. URL: https://www.gov.gd/index.php/economy.

(7) Bézy VS, Valverde RA, Plante CJ. Olive ridley sea turtle hatching success as a function of the microbial abundance in nest sand at Ostional, Costa Rica. PLoS One 2015;10(2):e0118579.

(8) Senko J, Nichols WJ, Ross JP, Willcox AS. To eat or not to eat an endangered species: Views of local residents and physicians on the safety of sea turtle consumption in northwestern Mexico. Ecohealth 2009;6(4):584-95.

(9) Zieger U, Trelease H, Winkler N, Mathew V, Sharma RN. Bacterial contamination of Leatherback turtle (Dermochelys coriacea) eggs and sand in nesting chambers at Levera Beach, Grenada, West Indies - a preliminary study. West Indian Vet J 2009;9(2):21-6.

(10) Aguirre AA, Gardner SC, Marsh JC, Delgado SG, Limpus CJ, Nichols WJ. Hazards associated with the consumption of sea turtle meat and eggs: A review for health care workers and the general public. Ecohealth 2006;3(3):141-53.

(11) Aguirre AA, Gardner SC, Marsh JC, Delgado SG, Limpus CJ, Nichols WJ. Potential human health risks associated with the consumption of sea turtle meat and eggs: A global perspective. Book Abstracts 2006;1:44.

(12) Corrente M, Sangiorgio G, Grandolfo E, Bodnar L, Catella C, Trotta A, et al. Risk for zoonotic salmonella transmission from pet reptiles: A survey on knowledge, attitudes and practices of reptile owners related to reptile husbandry. Prev Vet Med 2017;146(1):73-8.

(13) Barreiros JP, Barcelos J. 2001. Plastic ingestion by a leatherback turtle (Dermochelys coriacea) from the Azores (NE Atlantic). Marine Pollut Bulletin 2001;42(11):1196-7.

(14) George RH. Health problems and diseases of sea turtles. In: Lutz PL, Musick JA, eds. The biology of sea turtles. Washington, DC: CRC Press, 1996:363-85.
(15) Ives AK, Antaki E, Stewart KM, Francis S, Jay-Russell MT, Sithole F, et al. Detection of Salmonella enterica serovar Montevideo and Newport in free-ranging sea turtles and beach sand in the Caribbean and persistence in sand and seawater microcosms. Zoonoses Public Health 2017;64(6):450-9.
(16) Troëng S, Chacon D, Dick B. Possible decline in leatherback turtle Dermochelys coriacea nesting along the Caribbean Central America. Oryx 2004;38(4):395-403.

Chapter 14

Pandemic perceptions and attitudes towards health risk, adoption and avoidance behaviors

Susmitha Unni, MD/MPH
Sara Elzibak, BSc
and Satesh Bidaisee*, DVM, MSPH, EdD
Public Health and Preventive Medicine, St George's University, Grenada, West Indies

Abstract

The World Health Organization (WHO) declared SARSCoV2 a pandemic in March 2020. Since then, public health professionals have recommended hygiene-related and avoidance behaviors to curb the number of coronavirus infections. Through the online open access course 12, 234 participants enrolled from 120 countries on "An examination of Coronavirus COVID-19." The course provided the opportunity to understand the participants' perceptions and attitudes towards the pandemic. The study also aimed to explore the psychosocial and demographic factors associated with adopting the recommended hygiene-related and avoidance behaviors. Analysis revealed approximately 50% of the course's participants were 'somewhat' concerned about the risk of COVID-19 affecting their health and wellbeing. This evidently proved that there was only moderate concern of the infection which directly contributed towards the attitude and prevention behavior adopted. Through participants discussion posts, trust in government was one of the determining factors for the public to adopt the recommended hygiene-related and avoidance behaviors. Analysis

* ***Correspondence:*** Satesh Bidaisee, Professor, Public Health and Preventive Medicine, St George's University, Grenada, West Indies. Email: sbidaisee@sgu.edu

In: Public Health: Intersection of Health, Humans, Animals and the Environment
Editors: Satesh Bidaisee and Joav Merrick
ISBN: 979-8-89530-443-3

also suggests that an increase in science/health literacy levels was associated with increased adoption of behaviors. 84.9% of participants performed 1 of the 3 recommended hygiene related behaviors and 93.4% performed at least 1 of three avoidance-related behaviors. Social cognition and health belief model was applicable towards understanding people's perceptions and the rationale towards their attitudes and ultimately adopting a behavior to control the spread of COVID-19 infections.

Introduction

Our journey with Coronavirus began in late 2019 and has been an ongoing battle to date. The virus that was once unknown and then named 2019 nCoV was first identified by Huang et al., (1) in Wuhan in December 2019 and was published online in January 2020. All the reported cases were initially in the province of Wuhan in China. Until the first international case of 2019 nCoV was confirmed in Thailand at Bangkok airport (2). The second international case was also confirmed in Thailand (2). On January 21st, 2020, the first case of coronavirus was confirmed in Washington state, in the United States in a passenger who returned from Wuhan (3). The next confirmed outbreak was on March 6th, 2020, in a California cruise ship set to dock in San Francisco (4). Even with these outbreaks, there was a lot of uncertainty with the virus. There were no definitive answers for its genetic composition, etiology, mode of transmission, symptoms and treatment. On March 11th, 2020, WHO declared coronavirus infections a pandemic (5). Various social distancing, hygiene and quarantine practices were suggested by the WHO to curb the spread of this infection, which initially helped flatten the curve of incidence rates of the infection (6). Eventually on July 2nd, various states in the United States and other countries considered relaxing safety protocols and reopening public places, but were forced to revise their plans due to rising cases of infections again (6). By December 2020, the distribution of vaccinations to the general public was begun (7). Despite these vaccinations, there have been various strains like Delta and Omicron variants that have been causing failure against this battle with coronavirus. Currently, there have been 414,887,006 reported cases and 5,848,669 deaths due to COVID-19 worldwide. This has been an on-going battle for two years with waves of infections and re-infections. Hygiene-related and avoidance behaviors have been identified, yet the rates of infection remain high worldwide.

It is time to take a closer look at public health behaviors to determine the current state of this fight against coronavirus. It has been identified that health promotion is strongly dependent on the knowledge-attitude-behavior continuum (8). Providing people with knowledge about COVID-19 and conducting informational campaigns about the importance of hygiene-related behaviors like hand washing, sanitizing, and wearing a mask, as well as avoidance behaviors like social distancing, staying at home and quarantining will give people the opportunity to analyze their thoughts that can lead to attitude changes and subsequent desired behavior changes (8). However, high correspondence and appropriate measures are required to increase the magnitude of attitude-behavior relation (9). There are various factors that require attention to create a high correspondence environment. These include sources, mode of delivery and frequency of messages (8). In addition to knowledge, self-efficacy and cue to action are also important determinants to adopt hygiene-related and avoidance behaviors during COVID-19 (10).

Despite all the efforts to provide the most accurate information to the public, there are a lot of myths and misconceptions about the pandemic. There are also fatalistic beliefs that external forces control the infection, and humans have no power or influence over them (11). These myths, misconceptions and beliefs were common among low socio-economic groups, various ethnic backgrounds, and low health literacy population (12-14).

Given all our current knowledge of the coronavirus and our continued attempts to win over COVID-19, the ongoing pandemic is still claiming lives every day. The aim of this study is to examine the attitudes and perceptions of the population towards the COVID-19 pandemic, and the willingness and capacity of the general public to engage with mitigation measures. This study will explore the psychosocial and demographic factors that are associated with adoption of recommended hygiene-related and avoidance behaviors. The much-anticipated light at the end of the tunnel is close: when masks are allowed to be removed, public spaces re-open to full capacity and life is lived similarly to the pre-pandemic days.

Our study

During the first year of the pandemic, from March through the remainder of 2020, St George's University created an online course: An examination of Coronavirus COVID-19. This was an open access course available to participants worldwide. 12,234 participants were enrolled from 120 different countries. The course was advertised through an existing database of participants from the open access online course series. The participants were from different countries comprising the continents of Asia, Africa, North America, South America, Europe, and Australia.

The course was designed for participants to interact with experts and non-experts to discuss the novel coronavirus infection, review scientific evidence, analyze the global healthcare response to the pandemic and initiate intriguing discussions about the determinants of COVID-19 outbreak. The course was divided into four modules to integrate the different aspects of the infection and its global effects. Every module began with 1-2 questions to prompt thoughts and ideas for upcoming lectures and ended with a discussion prompt related to the module. Additionally, every module was supplemented with reading materials associated with the topics.

Participants were advised to attend lectures, answer questionnaires, and participate in discussions. Various discussion prompts were utilized to evaluate the perceptions and attitudes towards COVID-19 (Appendix-1). Live discussion panels with experts in their fields were also conducted. Participants were updated with situation reports of the COVID-19 pandemic. The four modules of the course were divided as follows:

Module 1- Global medical context of COVID-19: This module focused on discussing the emergence of a pandemic and the excess strain on medical professionals globally. A questionnaire was designed to learn about the concern of COVID-19 amongst the participants worldwide. The answers were measured on a scale of 1-10, 1 being least concerned and 10 being most concerned about COVID-19 affecting their health.

Module 2- Microbiology and transmission of COVID-19: To better understand the determinants of clinical care, prevention, and control strategies, this module focused on the biological composition of the virus, its route of transmission, the pathogenesis and diagnosis.

Module 3- Treatment and management: The focus of this module was to discuss the local and international preparedness for COVID-19 including diagnosis, treatment, hygiene precautions, health education and promotion. The chief medical officer of Grenada provided a situational analysis on the medical response in Grenada, and the Caribbean. A questionnaire was designed to determine the number of hygiene-related and avoidance behaviors adapted by the participants during the observed time. The answer choices included wash hands regularly, wear a mask, limit personal interaction, clean your workspace, telephone, doorknobs, etc.

Module 4- International health regulations for COVID-19: Prevention and control strategies for COVID-19 has been the primary target for every country worldwide. Through this module, participants had the opportunity to discuss the health policies and regulations put forth by the WHO. Participants were asked to rate their confidence in a unified global effort in response to COVID-19 and/or other similar emerging infectious disease, given the current geo-political and socio-economic structure of the world.

The course was designed to be self-paced; progression through modules in numerical order was not required. Participants were provided with a course completion certificate and thanked for their participation.

Findings

The question in Module 1 was: How would you rate your concern about COVID-19 affecting your personal health and well-being on a scale of 1 to 10 with one being the least concerned and 10 being the most concerned? Approximately 50% of the respondents reported 'somewhat' concern about COVID-19 and 19% were extremely concerned about COVID-19 affecting their personal health and well-being.

As per the questionnaires, 84.9% of participants performed ^{3} 1 of the three recommended hygiene related behaviors (see Figure 1). 93.4% performed ^{3} 1 of three avoidance-related behaviors (see Figure 2). The surveys and discussions throughout the course indicated that adopting hygiene and avoidance behaviors were associated with trust in the government. Another factor contributing to adopting mitigation measures is the science literacy level. The average rate of concern of COVID-19 was lowest in the North American continent, followed by Europe. The highest rate of concern was amongst Africa and South America (see Figure 2). The rate of concern also correlated with the number of hygiene related behaviors and avoidance behaviors adopted by the participants. Module 4 asked the participants: How confident are you, based on the current geo-political and socio-economic structure of the world, that a unified global effort can be achieved in response to COVID-19 and/or other similar emerging infectious diseases? Approximately 25.36% of the participants are completely confident that a unified global effort can be achieved. However, approximately 13.92% of the participants have no confidence at all (see Figure 3).

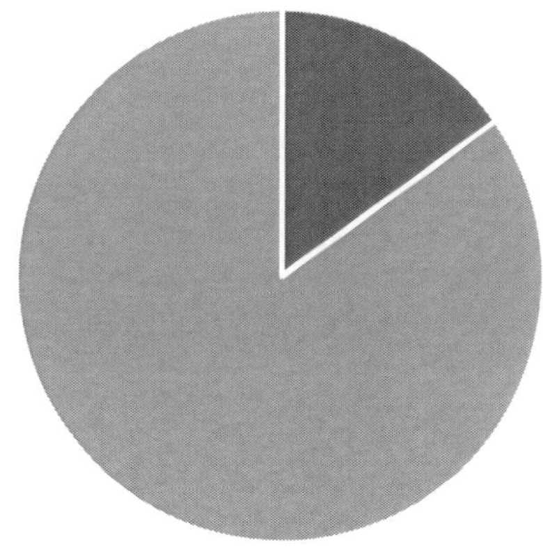

Figure 1. Number of recommended hygiene related behaviors performed by participants.

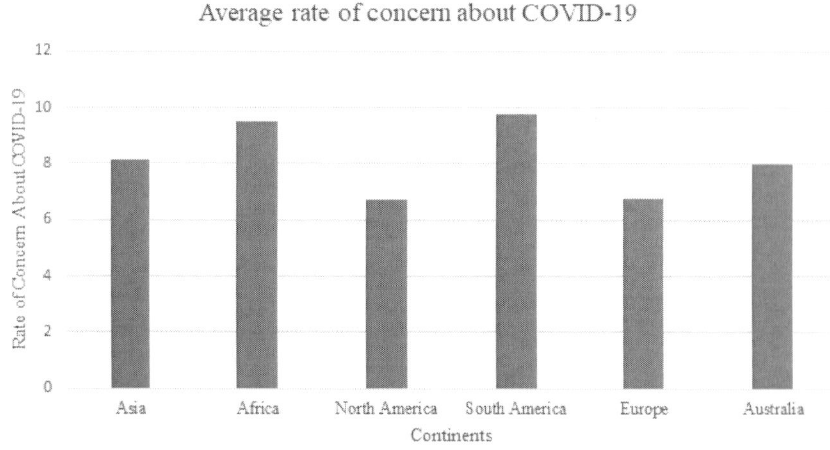

Figure 2. Average rate of concern among participants about COVID-19 affecting health and well-being.

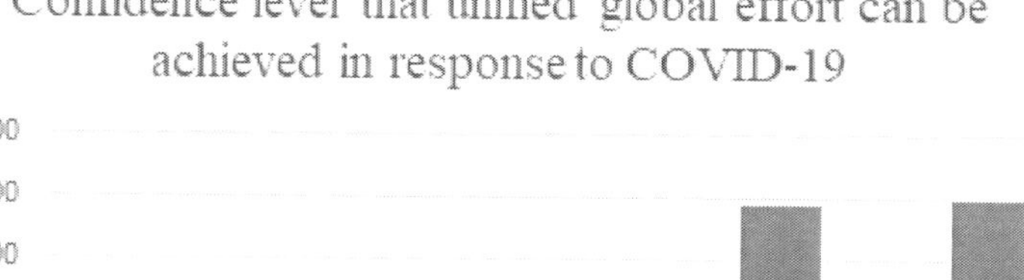

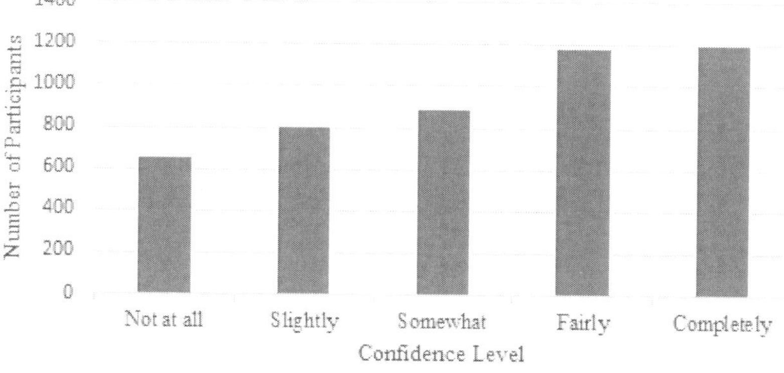

Figure 3. Confidence level that given the current geo-political and socio-economic structure of the world, that a unified global effort can be achieved in response to COVID-19 and/or other similar emerging infectious diseases.

Discussion

The social cognition model refers to a set of theories that state an individual's beliefs and attitudes are the determinants of their behavior (15). A subset of this model is the health belief model that was specifically developed for the field of public health to determine why certain people refuse vaccinations and screenings (15). This model describes that perceived risk of contracting the disease, perceived severity, benefits, and disadvantages play a role in adopting any recommended behaviors (15). The COVID-19 pandemic has highlighted this behavioral model worldwide. In countries with higher fear of susceptibility and effect of COVID-19, there were higher number of people adopting hygiene-related and avoidance behaviors. The discussion posts from the participants indicated that some of this fear of susceptibility and severity was inculcated by their respective governments sharing the appropriate information with the public. Therefore, it can be derived that there is a direct relationship between trusting the government and adopting these behaviors. A longitudinal study performed by Qing et al., (16), describes the increased rates of health behavior adoption associated to higher trust in the government. With government officials providing false reassurance and misinformation, it has created a lack of trust between the public and the politicians, which has caused detrimental effects worldwide (17). Some of the ways that governments can earn the trust of their people are through transparent communications, civic engagements and considering the needs of the community (18).

There are various other factors that contribute towards the adoption of mitigation behaviors; these include age, financial status, and mental health concerns (19). Participants in the course's discussion posts indicated that people from higher socio-economic backgrounds could sustain themselves and therefore participate in mitigation measures to reduce the spread of COVID-19 infections. The surveys and response to discussion posts also depicted a positive correlation between science literacy and adoption of hygiene-related and avoidance behaviors. The lack of scientific knowledge results in myths and misconceptions. Even though evidence-based information is provided to the public, there is still resistance to adopting preventative behaviors. Education is the cornerstone to developing self-efficacy, which is the most important determinant of preventative behaviors (20). Health communication professionals will have to implement education and awareness communication styles tailored specifically to each group of resistant population (21). WHO has implemented a special platform for risk communication in 2020 called WHO Information Network for Epidemic (EPI-WIN), to work faster than the speed of unverified news sweeping through social media to inform and educate people appropriately with evidence-based recommendations (22, 23). WHO also initiated a website called "myth busters" in attempts to advise the public about COVID-19 (24).

Social psychology has described the relationship between attitudes and behaviors through two models, a spontaneous and a deliberative processing model of the attitude-behavior relation. The spontaneous behavior model depends on a previously developed attitude from memory, while the deliberative processing model describes that humans build attitudes based on an analysis of the positive and negative features of the attitude object, as well as the costs and benefits (25). After careful reflection and weighing the course of action, it leads to forming a behavior intention, and finally behavior (19). If governments are able to provide the public with the required information and support to deliberately analyze the course of their actions, it can provide for increased adherence to hygiene-related and avoidance behaviors. It will also help create an attitude memory towards the object that can eventually lead to preventative spontaneous behaviors.

Through the observed time of the course with training through the various models, it helped increase the level of knowledge and science literacy, thus providing an opportunity for the participants to analyze their attitudes. This eventually translated into increased number of participants reporting higher adoption of safety and mitigation practices. During this pandemic, individuals with perceptions of high susceptibility, high severity, high benefits, and low barriers are most likely to adopt the recommended hygiene-related and mitigation behaviors (26). Since this course was delivered virtually, there were a few limitations to it. Only people with internet access and familiarity to digital information were able to participate in this course. Additionally, the age group for participation was limited, and did not include children and elderly, due to accessibility to the online forum of this course.

Conclusion

Given the complexity of the human mind, it is without doubt that appropriate information must be provided to individuals for effortful analysis to build attributes and deliberately process the information to perform the behavior. Public health officials through the government and other science literacy avenues can inform and educate the public about COVID-19. This will provide people with all the information required to make rational decisions and adopt the recommended safety protocol and mitigation measures. It is imperative to understand and create a targeted approach towards people's perceptions and attitudes about the disease, as this will eventually lead to adopting the recommended behaviors. It is critical to install and sustain these behaviors to curb the spread of COVID-19 infections worldwide. Given the current universal progression of COVID-19, a further analysis into maintenance of hygiene-related and avoidance behaviors can provide further insight into the perceptions and attitudes of the COVID-19 pandemic.

References

(1) Huang C, Wang Y, Li X, Ren L, Zhao J, Hu Y, et al. Clinical features of patients infected with 2019 novel coronavirus in Wuhan, China. Lancet 2020;395(10223):497-506.

(2) Allam Z. The first 50 days of COVID-19: A detailed chronological timeline and extensive review of literature documenting the pandemic. In: Allam Z. Surveying the COVID-19 pandemic and its implications. Victoria, Australia: Elsevier, 2020:1-7.

(3) Centers for Disease Control and Prevention. First travel-related case of 2019 novel coronavirus detected in United States. Atlanta, GA: CDC, 2020. URL:

https://www.cdc.gov/media/releases/2020/p0121-novel-coronavirus-travel-case.html.

(4) AJMC. COVID-19 roundup: 21 passengers on California cruise ship test positive. AJMC 2020. URL: https://www.ajmc.com/view/covid19-roundup.

(5) AJMC. COVID-19 roundup: Coronavirus now a national emergency, with plans to increase testing. AJMC 2020 Mar 07. URL: https://www.ajmc.com/view/covid19-roundup2.

(6) AJMC. What we're reading: Covid-19 reinfection in us; remdesivir use expanded; colon cancer screening. AJMC 2020 Aug 31. URL: https://www.ajmc.com/view/what-we-re-reading-covid-19-reinfection-in-us-remdesivir-use-expanded-colon-cancer-screening.

(7) AJMC. FDA advisory panel recommends Pfizer, BioNTech Covid-19 vaccine. AJMC 2020 Dec 11.

(8) Bettinghaus EP. Health promotion and the knowledge-attitude-behavior continuum. Prevent Med 1986;15:475-91.

(9) Ajzen I. Attitude-behavior relations: A theoretical analysis and review of empirical research. Psychol Bull 1977;84(5):888-918.

(10) Lv G, Yuan J, Hsieh S, Shao R, Li M. Knowledge and determinants of behavioral responses to the pandemic of COVID-19. Front Med 2021;8:673187. https://www.ncbi.nlm.nih.gov/pmc/articles/PMC8219873/

(11) Shahnazi H, Ahmadi-Livani M, Pahlavanzadeh B, Rajabi A, Hamrah MS, Charkazi A. Assessing preventive health behaviors from COVID-19: A cross sectional study with health belief model in Golestan Province, Northern of Iran. Infect Dis Poverty 2020; 9(6):91-9.

(12) Bin Naeem S, Boulos MNK. COVID-19 misinformation online and health literacy: A brief overview. Int J Environ Res Public Health 2021;18:8091.

(13) Mistry SK, Ali ARMM, Yadav UN, Ghimire S, Hossain B, Saha M, et al. Misconceptions about COVID-19 among older Rohingya (forcefully displaced Myanmar nationals) adults in Bangladesh: Findings from a cross-sectional study. BMJ Open 2021;11(5):e050427.

(14) Bakebillah M, Billah MA, Wubishet BL, Khan MN. Community's misconception about COVID-19 and its associated factors in Satkhira, Bangladesh: A cross-sectional study. PLoS One 2021;16(9):e0257410.

(15) Shahnazi H, Ahmadi-Livani M, Pahlavanzadeh B, Rajabi A, Hamrah MS, Charkazi A. Assessing preventive health behaviors from COVID-19: a cross sectional study with health belief model in Golestan Province, Northern of Iran. Infectious diseases of poverty. 2020;9(06):91-9.

(16) Han Q, Zheng B, Cristea M, Agostini M, Belanger J, Gutzkow B, et al. Trust in government regarding COVID-19 and its associations with preventive health behaviour and prosocial behaviour during the pandemic: a cross-sectional and longitudinal study. Psychol Med 2021;1-11.

(17) Larson H. Blocking information on COVID-19 can fuel the spread of misinformation. Nature 2020 Mar 30. URL: https://www.nature.com/articles/d41586-020-00920-w.

(18) Hyland-Wood B, Gardner J, Leask J, Ecker UKH. Toward effective government communication strategies in the era of COVID-19. Humanit Soc Sci Commun 2021;8(30).
(19) Myerson J, Strube MJ, Green L, Hale S. Individual differences in COVID-19 mitigation behaviors: The roles of age, gender, psychological state, and financial status. PLoS One 2021;16(9):e0257658.
(20) Bayat F, Shojaeezadeh D, Baikpour M, Heshmat R, Baikpour M, Hosseini M. The effects of education based on extended health belief model in type 2 diabetic patients: a randomized controlled trial. J Diabetes Metab Disord 2013;12(1):45.
(21) Resnicow K, Bacon E, Yang P, Hawley S, Van Horn ML, An L. Novel predictors of COVID-19 protective behaviors among US adults: Cross-sectional survey. J Med Internet Res 2021;23(4):e23488.
(22) Zarocostas J. How to fight an infodemic. Lancet 2020;395(10225):676.
(23) Islam AKMN, Laato S, Talukder S, Sutinen E. Misinformation sharing and social media fatigue during COVID-19: An affordance and cognitive load perspective. Technol Forecast Soc Change 2020;159:120201.
(24) World Health Organization. How to report misinformation. Geneva: WHO, 2022. URL: https://www.who.int/emergencies/diseases/novel-coronavirus-2019/advice-for-public/myth-busters.
(25) Fazio R. Multiple processes by which attitudes guide behavior: The mode model as an integrative framework. Adv Experiment Soc Psychol 1990;23:75-109.
(26) Becker M. The health belief model and sick role behavior. Health Educat Monogr 1974;2(4):409-19.

Appendix-1. Discussion prompt questions for every module during the online course: An Examination of Coronavirus COVID-19

Module	Discussion prompt
Global Medical Context of COVID-19	Diseases including COVID-19 is a result/consequence of environmental and other changes which would have interacted to allow for its emergence. In the case for COVID-19, what are some of these predisposing factors which contributed to its emergence?
Microbiology and Transmission of COVID-19	How does the COVID-19 virus structure predict its distribution and projected longevity in the global human population?
Treatment and Management of COVID-19	A key strategy in managing COVID-19 is the promotion of hygiene and sanitation practices. With advances in medicine and society, why does sanitation and hygiene remain inadequate?
International Health Regulations for COVID-19	Consider existing protocols for the global response to COVID-19. Do you think there is a need to review and revise the current strategy? Why or why not?

Section three: Acknowledgments

Chapter 15

About the editors

Satesh Bidaisee, DVM, MSPH, EdD, was born in Trinidad and had an early exposure to natural sciences through his family background. An early and sustained passion for life especially nature fostered a childhood and schooling in the biological sciences. Satesh through trials and tribulations now serves as a Professor of Public Health and Preventive Medicine and Associate Dean for Graduate Studies at St George's University. He has previously held positions at the University of Trinidad and Tobago and the Ministry of Health in Trinidad and Tobago. Satesh is a graduate of the University of the West Indies, Faculty of Medical Sciences, St Augustine, Trinidad, St George's University, School of Medicine, School of Graduate Studies and the University of Sheffield, United Kingdom. As a research investigator, Satesh supports community based participatory research and service activities along his interests which include emerging infectious diseases, zoonoses, food safety and food security and the pursuit of the One Health One Medicine concept. Satesh's research projects include human behavior, climate change and viral infections, zoonoses and One Health and vector borne disease outbreak investigation. Satesh is board certified by the United States National Board of Public Health Examiners, and holds Fellowships to the Royal Society of Public Health (FRSPH), Royal Society of Tropical Medicine and Hygiene (FRSTMH), International Society on Infectious Diseases and the Society of Biology. Email: sbidaisee@sgu.edu

Joav Merrick, MD, MMedSci, DMSc, born and educated in Denmark is a professor of pediatrics, Division of Pediatrics, Hadassah Hebrew University Medical Center, Mt. Scopus Campus, Jerusalem, Israel and Kentucky Children's Hospital, University of Kentucky, Lexington, Kentucky United States and professor of public health at the Center for Healthy Development, School of Public Health, Georgia State University, Atlanta, United States, the former medical director of the Health Services, Division for Intellectual and Developmental Disabilities, Ministry of Social Affairs and Social Services, Jerusalem, the founder and director of the National Institute of Child Health and Human Development in Israel. Numerous publications in the field of pediatrics, child health and human development, rehabilitation, intellectual disability, disability, health, welfare, abuse, advocacy, quality of life and prevention. Received the Peter Sabroe Child Award for outstanding work on behalf of Danish Children in 1985 and the International LEGO-Prize ("The Children's Nobel Prize") for an extraordinary contribution towards improvement in child welfare and well-being in 1987. In 2017 appointed a Kentucky Colonel by the Commonwealth of Kentucky, the highest honor the governor can bestow to a person. Email: jmerrick@zahav.net.il

Chapter 16

About St George's University, Department of Public Health and Preventive Medicine (SGU-DPHPM), St George's, Grenada

Department of Public Health and Preventive Medicine (SGU-DPHPM) was established in 1999 to develop and deliver its Master of Public Health (MPH) graduate program. SGU as an international center of medical education welcomed the opportunity to train medical students and graduates in public health towards strengthening primary health care and serve capacity building for the Caribbean region towards its health systems strengthening. Since 1999, SGU-DPHPM has graduated over one thousand healthcare professionals that serve in over 50 countries. The mission of SGU-DPHPM is to provide an academic focal point for interdisciplinary education, research, service and scholarship with a focus that is both globally relevant and locally applicable.

Education

SGU-DPHPM administers a Council on Education for Public Health (CEPH) MPH program with areas of concentration in global health, preventive medicine, veterinary public health, epidemiology, environmental and occupational health and health policy and administration. SGU-DPHPM also coordinates public health education for the Doctor of Medicine (MD) program as well as collaborates in veterinary public health for the Doctor of Veterinary Medicine (DVM) program. Additionally, SGU-DPHPM provides graduate course for students in the School of Graduate Studies Masters and Doctoral programs, a Certificate in Public Health program for public health practitioners and supports workforce development and community health education through ongoing training initiatives. Students and graduates of SGU-DPHPM MPH program successfully enhance their graduate public health experience with a commitment to lifelong learning as Certified Public Health (CPH) professionals from the United States National Board of Public Health Examiners (NBPHE).

Research activities

The affiliated faculty of SGU-DPHPM over the years have published work from projects and research activities thorough national and international collaborations. For the past 20 years, in collaboration with SGU's research institute, Windward Islands Research and Education Foundation (WINDREF), SGU-DPHPM faculty, students and project/research partners and collaborators have administered over 10 million USD in grant funding and over 250 publications across peer reviewed journals and national and international publications. SGU-DPHPM research is supported by SGU's Office of Research which includes access to institutional funding and support as well as Institutional Review Board (IRB) and Institutional Animal Care and Use Committee (IACUC) which serves to review research work for ethical clearance. SGU-DPHPM research is coordinated with the research institute Windward Islands Research and Education Foundation (WINDREF). WINDREF, established in 1994 is registered as a 501(c)3 in the United States, a Charitable Trust in the United Kingdom and a Non-Governmental Organization in Grenada. Through WINDREF's coordination with administrative support, laboratory capacity, grants management, statistical support and network of research fellows, SGU-DPHPM faculty and students are actively engaged in research activities.

Service

Over the past 20th years of SGU-DPHPM, many activities became focused on collaboration with various local, regional and international initiatives, which has resulted in collaboration around the establishment of a World Health Organization (WHO) Collaborating Center in Environmental and Occupational Health. Also, in response to the disproportionate and adverse effects of the climate on health, a United Nations Frame Convention on Climate Change, Regional Collaborating Center (UNFCCC-RCC). Furthermore, SGU-DPHPM has partnered with local and regional governments and community based organizations in response to disasters, vector borne disease outbreaks and overall management of non-communicable diseases. Service activities is additionally conducted with SGU-DPHPM's active Public Health Students Association (PHSA). PHSA routinely partners with medical and veterinary students to conduct community health clinics as well as health education and environmental management service based activities across Grenada. SGU-DPHPM additionally contains the Gamma Kappa chapter of Delta Omega Honors Society which is the oldest professional public health society which prides itself on community service and informs community work in response to community assessment and identified community needs.

Scholarship

SGU-DPHPM focuses on inter-professional education, research and service is lead through the scholarship of its faculty and students collaborations and in partnership with international institutions and colleagues. Details of SGU-DPHPM's scholarship can be accessed from the website (sgu.edu).

Partners and collaborations

SGU-DPHPM education, research, service and scholarship is delivered in collaborations with the Schools of Medicine, Veterinary Medicine, Arts and Sciences and Graduate Studies at SGU. SGU-DPHPM also supports local organizations in Grenada with their annual work plan which includes the Grenada Public Health Association, Grenada Diabetes Association and the Grenada Cancer Society. Regionally, SGU-DPHPM supports the work of the Organization of Eastern Caribbean States (OECS), Caribbean Public Health Agency (CARPHA) and the Pan American Health Organization (PAHO). SGU-DPHPM through memberships and fellowships are also actively involved in the work of the American Public Health Association (APHA), Association of Schools and Programs in Public Health (ASPPH), Consortium on Universities in Public Health (CUGH), Royal Society of Public Health (RSPH), Royal Society of Tropical Medicine and Hygiene (RSTMH) and global agencies such as World Health Organization (WHO), World Bank and the United Nations (UN). Funding partnership and collaborations is engaged through agencies such as the National Institutes of Health (NIH), World Bank, World Diabetes Foundation (WDF), United Nations Development Program (UNDP) United Nations Global Environmental Facility (UN-GEF), Centers for Disease Control and Prevention (CDC), National Institutes for Occupational Safety and Health (NIOSH), United States Department of

Agriculture (USDA), Public Health Canada, African Medical Research and Education Foundation (AMREF) and the Caribbean Community (CARICOM) and Caribbean Public Health Agency (CARPHA). SGU-DPHPM with its interdisciplinary local to global focus have also partners with several institutions in the Caribbean; T.A. Marryshaw Community College (TAMCC), University of the West Indies, University of Trinidad and Tobago, University of Technology, Jamaica, Pasteur Institute, Guadeloupe, Institute of Tropical Medicine, Cuba, North America; New York University (NYU), State University of New York (SUNY), Sanford University, John's Hopkins University Bloomberg School of Public Health (JHU), Colombia University, Mailman School of Public Health, South America; Universidad de Antioquia, Colombia, University of Guyana, Guyana, Europe; Northumbria University (NU) and the Royal Veterinary College (RVC), UK, Swiss Tropical Medicine Institute, Switzerland, Africa; Juba Medical College, South Sudan, University of Kwa-Zulu, Natal, South Africa, Makerere University, Uganda, Asia; Mahidol University and Kasetsart University, Thailand, Chitkara University, MS Ramaiah Medical College and Indian Institute of Health Management Research (IIHMR), India, Konkuk University, South Korea and Polytechnic Institute in Singapore.

Contact

Satesh Bidaisee, EdD
Professor, Public Health and Preventive Medicine, St George's University
Grenada, West Indies
Email: sbidaisee@sgu.edu

Chapter 17

About the National Institute of Child Health and Human Development in Israel

The National Institute of Child Health and Human Development (NICHD) in Israel was established in 1998 as a virtual institute under the auspicies of the Medical Director, Ministry of Social Affairs and Social Services in order to function as the research arm for the Office of the Medical Director. In 1998 the National Council for Child Health and Pediatrics, Ministry of Health and in 1999 the Director General and Deputy Director General of the Ministry of Health endorsed the establishment of the NICHD.

Mission

The mission of a National Institute for Child Health and Human Development in Israel is to provide an academic focal point for the scholarly interdisciplinary study of child life, health, public health, welfare, disability, rehabilitation, intellectual disability and related aspects of human development. This mission includes research, teaching, clinical work, information and public service activities in the field of child health and human development.

Service and academic activities

Over the years many activities became focused in the south of Israel due to collaboration with various professionals at the Faculty of Health Sciences (FOHS) at the Ben Gurion University of the Negev (BGU). Since 2000 an affiliation with the Zusman Child Development Center at the Pediatric Division of Soroka University Medical Center has resulted in collaboration around the establishment of the Down Syndrome Clinic at that center. In 2002 a full course on "Disability" was established at the Recanati School for Allied Professions in the Community, FOHS, BGU and in 2005 collaboration was started with the Primary Care Unit of the faculty and disability became part of the master of public health course on "Children and society." In the academic year 2005-2006 a one semester course on "Aging with disability" was started as part of the master of science program in gerontology in our collaboration with the Center for Multidisciplinary Research in Aging. In 2010 collaborations with the Division of Pediatrics, Hadassah Hebrew University Medical Center, Jerusalem, Israel around the National Down Syndrome Center and teaching students and residents about intellectual and developmental disabilities as part of their training at this campus.

Research activities

The affiliated staff have over the years published work from projects and research activities in this national and international collaboration. In the year 2000 the International Journal of Adolescent Medicine and Health and in 2005 the International Journal on Disability and Human Development of De Gruyter Publishing House (Berlin and New York) were affiliated with the National Institute of Child Health and Human Development. From 2008 also the International Journal of Child Health and Human Development (Nova Science, New York), the International Journal of Child and Adolescent Health (Nova Science) and the Journal of Pain Management (Nova Science) affiliated and from 2009 the International Public Health Journal (Nova Science) and Journal of Alternative Medicine Research (Nova Science). All peer-reviewed international journals.

National collaborations

Nationally the NICHD works in collaboration with the Faculty of Health Sciences, Ben Gurion University of the Negev; Department of Physical Therapy, Sackler School of Medicine, Tel Aviv University; Autism Center, Assaf HaRofeh Medical Center; National Rett and PKU Centers at Chaim Sheba Medical Center, Tel HaShomer; Department of Physiotherapy, Haifa University; Department of Education, Bar Ilan University, Ramat Gan, Faculty of Social Sciences and Health Sciences; College of Judea and Samaria in Ariel and in 2011 affiliation with Center for Pediatric Chronic Diseases and National Center for Down Syndrome, Department of Pediatrics, Hadassah Hebrew University Medical Center, Mount Scopus Campus, Jerusalem.

International collaborations

Internationally with the Department of Disability and Human Development, College of Applied Health Sciences, University of Illinois at Chicago; Strong Center for Developmental Disabilities, Golisano Children's Hospital at Strong, University of Rochester School of Medicine and Dentistry, New York; Centre on Intellectual Disabilities, University of Albany, New York; Centre for Chronic Disease Prevention and Control, Health Canada, Ottawa; Chandler Medical Center and Children's Hospital, Kentucky Children's Hospital, Section of Adolescent Medicine, University of Kentucky, Lexington; Chronic Disease Prevention and Control Research Center, Baylor College of Medicine, Houston, Texas; Division of Neuroscience, Department of Psychiatry, Columbia University, New York; Institute for the Study of Disadvantage and Disability, Atlanta; Center for Autism and Related Disorders, Department Psychiatry, Children's Hospital Boston, Boston; Department of Pediatric and Adolescent Medicine, Western Michigan University Homer Stryker MD School of Medicine, Kalamazoo, Michigan, United States; Department of Paediatrics, Child Health and Adolescent Medicine, Children's Hospital at Westmead, Westmead, Australia; International Centre for the Study of Occupational and Mental Health, Düsseldorf, Germany; Centre for Advanced Studies in Nursing, Department of General Practice and Primary Care, University of Aberdeen, Aberdeen, United Kingdom; Quality of Life Research Center, Copenhagen, Denmark; Nordic School of Public Health, Gottenburg, Sweden, Scandinavian Institute of Quality of Working Life, Oslo, Norway; The Department of Applied Social Sciences (APSS) of The Hong Kong Polytechnic University Hong Kong.

Targets

Our focus is on research, international collaborations, clinical work, teaching and policy in health, disability and human development and to establish the NICHD as a permanent institute in Israel in order to conduct model research and provide policy aspects in the field of interest.

Contact

Professor Joav Merrick, MD, MMedSci, DMSc

Director, National Institute of Child Health and Human Development, Jerusalem, Israel. E-mail: jmerrick@zahav.net.il

Section four: Index

Index

A

B

C

D

M

N

O

P

R

S

T

U

V

W

Z